CAMARO PERFORMANCE
HANDBOOK

by David Shelby

HPBooks

ACKNOWLEDGMENTS

There are many people whose help was invaluable in writing this book. It is with my sincerest gratitude that I thank each and every one of them for their assistance or input in making *Camaro Performance* a reality. But the most thanks must go to my wife Elizabeth who tolerated me, my work, my mess, my time at the race track, the time devoted to my Camaro and this book.

Rick Navaro, Pacific Performance
Kelly Irwin, Innovative Imaging
Don Alexander, Suspension Expert
Daniel Sanchez, All Chevy Magazine
Mike Lutfy, HPBooks
Dave Emanuel, Performance Publishing Co.
J.R. Granatelli, Paxton Products
Bob Brennon, Classic Camaro
Suzanne Kane, Chevrolet Public Relations
Kay Ward, Chevrolet Public Relations
Tom Hoxie, Chevrolet Public Relations
Ralph Kramer, Chevrolet Public Relations
Mark MacPhail, Chevrolet Race Shop
Chuck Maguire, GM Powertrain
Rod Michaelson, Chevrolet Engineering
Los Angeles County Raceway
Bill Huth, Willow Springs Raceway
Glenn Grissom, Circle Track Magazine
Joe Granatelli, Granatelli Performance
Ron Zimmerman, Z Industries
Jeff Cheechov, Suspension Techniques
Bill Jonbloed, Jonbloed Wheels
Willie Robinson, Classic Sixties
Sylvia Welch, Turbo City
Camee Edelbrock, Edelbrock Corp.
Mike Vendetto, Accel
Kerry Novack, Crower Cams & Equipment
Greg Hinkfus, JFZ

HPBooks
Published by The Berkley Publishing Group
A division of Penguin Putnam Inc.
375 Hudson Street, New York, New York 10014
© 1993 Price Stern Sloan, Inc.

Printed in the United States of America

20 19 18 17 16 15 14 13 12 11 10 9 8

The Penguin Putnam Inc. World Wide Web site address is
http://www.penguinputnam.com

Library of Congress Cataloging-in-Publication Data

Shelby, David.
   Camaro Performance / by David Shelby.
   p. cm.
   Includes Index
   ISBN 1-55788-057-3
   1. Camaro automobile—Performance. I. Title.
TL215.C33S49 1993
629.222´2—dc20   93-14277
                                                                 CIP

NOTICE: The information in this book is true and complete to the best of our knowledge. All recommendations on parts and procedures are made without any guarantees on the part of the author or the publisher. Tampering with, removing or altering any emissions-control device is a violation of federal regulations. Author and publisher disclaim all liability incurred in connection with the use of this information.

Cover Photo: Courtesy *All Chevy* magazine.
Interior Photo & Illustrations: David Shelby; others noted
Production: David Bird and Eric Van Eyke

All rights reserved. No part of this work may be reproduced or transmitted in any form by any means, electronic or mechanical, including photocopying and recording, or by any information storage or retrieval system, without written permission from the publisher, except in the case of brief quotations embodied in critical articles or reviews.

This book is printed on acid-free paper.

CONTENTS

Chapter One: A Brief History 1

Chapter Two: Computerized Engine Control Systems 15

Chapter Three: Ignition Systems 29

Chapter Four: Exhaust Systems 39

Chapter Five: Induction Systems 53

Chapter Six: Cylinder Heads & Valvetrain Modifications .. 67

Chapter Seven: Forced Induction 79

Chapter Eight: Drivetrain 93

Chapter Nine: Suspension 107

Chapter Ten: Interior Modifications 125

Index 137

The third-generation Camaro debuted with high expectations of performance and styling among the media and car buffs alike. It didn't let them down.

THE THIRD GENERATION

A BRIEF HISTORY

The Chevrolet Camaro has evolved through four generations since its introduction on September 12, 1966. Since that time, the Camaro has always represented a tremendous performance value. With each major change, the Camaro just seemed to get better, and of course there were plenty of aftermarket parts available to add just about any level of performance you wanted. Now, with the January 1993 introduction of the fourth-generation Camaro, the 11-year reign of the highly successful third-generation models now comes to an end—or should we say beginning? The 1982 to 1992 Camaro models continue to enjoy exceptional popularity amongst performance enthusiasts, and the aftermarket has responded with numerous performance items that are relatively easy to install, and many are emissions-legal. Each of the 11 model years offered increased levels of performance, all that lend themselves to modern day hot rodding.

The 1982 Camaro was also chosen as the Indianapolis 500 Pace Car for the third time in Camaro history. Shown here is one of the total 6360 Commemorative Edition Pace Cars built.

Here's a breakdown of the technical highlights for each model year.

1982

When the third-generation Camaro was about to debut for the 1982 model year, Chevrolet was trying to maintain the Camaro's performance appeal by continuing the Z28 package in an all-new body style. However, that didn't prevent a lot of second-guessing, about body design and powertrains, on the part of the automotive press and die-hard Chevrolet enthusiasts. Many "knowledgeable" automotive journalists had predicted that the third-generation Camaro would be a front-wheel-drive vehicle that would not live up to the per-formance appeal of previous models.

Chevrolet designers had other ideas. Although a front-wheel-drive platform was seriously considered, the ultimate decision-makers felt that the most important aspect of the '82 Camaro was retention of a linkage to its performance heritage (over a decade later, the designers of the fourth-generation Camaro thought the same thing). That meant one thing—a front-mounted V8 turning rear drive wheels. And to add some icing on the cake, Chevrolet molded the third-generation Camaro into a lighter, sleeker and more stylish vehicle than its ancestor. In fact, the 1982 Camaro was very close in size and weight to the very first Camaro that hit the streets back in 1967.

Pace Car Edition—To say the least, the automotive media and the industry in general were very impressed by the 1982 Camaro. It was selected as *Motor Trend* magazine's *Car of the Year* and was also chosen as the Official Pace Car for the 1982 Indianapolis 500. This was the third of four times in the Camaro's history that it had been chosen to pace the field at Indy (1967, 1969 and 1993 are the other three occasions).

As with the previous Camaro pace cars, Chevrolet produced a limited number of pace car replicas. A total of 6,360 "Commemorative Edition" 1982 Camaro pace cars were produced. These replicas featured a 305 cubic-inch (5-liter) V8, silver and blue paint with Indy 500 logos on the doors, sport mirrors, aluminum wheels and Goodyear Eagle GT tires.

Z28—The Z28 package for 1982 included a base small-block 305 V8 which was rated at 145 horsepower. This engine featured standard cast-iron heads with a hydraulic camshaft, which had a mild profile: 202 intake/207 exhaust duration at .050" lift and .404" intake/.415" exhaust gross valve lift. The engine's compression ratio was 9.3:1, giving the 305 adequate power but with good fuel economy.

As for transmissions in 1982, a Borg-Warner Super T-10 4-speed was mated to V8 engines. This transmission included a special tailhousing

The 1983 models had several engine options including one with the ill-fated Cross Fire Injection. The carbureted 5.0 Liter H.O. proved to be a much more popular option.

that had a provision to mount the torque bar that extended forward from the rear axle. Also available was a Turbo Hydra-matic 200 three-speed automatic transmission.

The 1982 model year also brought a new type of electronic fuel injection called Throttle Body Injection (TBI) to the Camaro's engine bay. This was available as part of Regular Production Option (RPO) LU5, which was a 165 horsepower version of the 305 small block.

Another feature introduced with the 1982 model was MacPherson strut front suspension. Previous generations had been fitted with an SLA (Short and Long Arm) suspension similar in design to the one used on 1982-and-earlier Corvettes. The switch to front struts was allegedly instigated by a number of engineering considerations (lighter weight and lower front-end profile) but in retrospect, it appears to have been a marketing decision. Front struts were the hot trend in the late 1970s and early 1980s when the third-generation Camaro was designed. However, although they offer a potential cost savings, the geometry of a MacPherson strut suspension is less than ideal for superior ride and handling. Testimony to that fact can be found in the front suspension of the fourth-generation Camaro which is of the SLA persuasion, and offers vastly improved ride and handling characteristics over the predecessor.

The Berlinetta Camaros for 1984 had a unique digital speedometer and swivel radio.

The year 1985 was the first for the IROC-Z but the standard Z28 package was also available. The IROC-Z had stiffer suspension, a 305 cid engine with Tuned Port Injection and special graphics.

1983

For the 1983 model year, Chevrolet maintained the course established the previous year. Horsepower of the standard 305 V8 was bumped to 150 and two optional engine packages were also available. One included electronically controlled Cross Fire Injection which was the standard induction system on the Corvette. However, the Cross Fire quickly lost much of its popularity as it was prone to water and vacuum leaks from improper sealing. Although it was soon discontinued, it did raise the output of 305 cid engines to 175 horsepower at 4200 rpm and 250 lbs-ft. of torque at 2800 rpm. The other engine option, which became the choice of the majority of performance-oriented Camaro buyers, was a 190 horsepower 305 cubic-inch, High Output engine which featured an electronically controlled Rochester QuadraJet carburetor, a performance camshaft with higher lift and longer duration and low restriction intake and exhaust systems. This option, known as RPO L69, cost the same as the more exotic Cross Fire system, yet resulted in more horsepower. Another strike against the Cross Fire system was a lack of willingness, on the part of performance enthusiasts, to experiment with a computer-controlled electronic fuel injection system. Even though carburetors were old news, the H.O. option was the more popular choice—in spite of the fact its carburetor was computer controlled.

Two new transmissions were also available in the Berlinetta and Z28 in 1983; the 700R4 four-speed automatic and the Borg-Warner T-5 five-speed manual gearbox.

1984

After two years of great success, Chevrolet continued to improve the Camaro for the 1984 model year. Standard equipment on the Berlinetta included a digital instrument panel, featuring a speedometer display and bar-type tachometer. It was also the only model to feature a center console mounted radio on a swivel pod, and a roof console that contained a map light, removable flashlight and mileage counter spools. This roof console was also available on other Camaro models under RPO DK6. Collectors should note that the roof console option on 1984 Z28 models is rare and adds some value.

In 1984, the Camaro again made the news as it was chosen by *Road & Track* magazine as one of the twelve best cars in the world (under the sports GT category in its price range). *Car and Driver* magazine also named the '84 Camaro Z28 as the "best handling car built in the United States," choosing it over the Ford Mustang, Pontiac Trans Am and even the Camaro's more expensive half-brother, the Corvette.

Interior changes included the addition of a new high-end stereo system, featuring an AM/FM cassette receiver with a built-in power booster and five-band equalizer. This system was the forerunner of future great stereo systems to come.

V8 powerplant selection for 1984 included the 150-horsepower LG4 and 190-horsepower L69 (305 H.O.) both of which were equipped with electronically controlled Rochester four-barrel carburetors.

About the only significant driveline change for 1984 was hydraulic clutch linkage in place of the conventional mechanical linkage used previously. Chevrolet engineers, however, had plenty of technical improvements for the following year.

1985

The word IROC would become synonymous with the Camaro in 1985, since the International Race of Champions racing series used the Camaro as its official vehicle. The IROC-Z's performance on the race track led the way for surprisingly effective improvements in the suspensions of standard production Camaros. As a result, handling improved noticeably.

IROC Features—In 1985, the IROC-Z Sport Equipment Package was available only on Z28 and was listed as RPO B4Z on the option sheet. IROC-Zs had a lower ride height than the standard Camaro and featured performance-calibrated front struts and springs, Bilstein rear shocks and 16-inch aluminum wheels with Goodyear Eagle 245/50/16 tires. The IROC-Z definitely had exceptional handling capabilities, recording skid pad numbers in the .89 g range. In conjunction with a new Tuned Port Injected 305 V8, the improved handling made the IROC-Z one of the hottest cars to set tire to pavement. The TPI-equipped 305 churned out 215 horsepower, and could be ordered only with a four-speed automatic transmission. The other option available for the IROC-Z was the 305 H.O. engine, which produced 190 horsepower and was backed by either a four-speed automatic or five-speed manual transmission.

For Camaro enthusiasts with a budget that couldn't handle the IROC-Z price tag, Berlinetta models featured a 173 cid (2.8-liter) V6 with a new Multi-Port Fuel Injection (MFI) system. For the most part, both the Berlinetta and the IROC came equipped with split, fold-down rear seats and a new instrument panel that eliminated the double-needle speedometer. The Berlinetta, however, continued to have the roof console as standard equipment.

1986

For the most part, the body and suspension remained the same for the IROC-Z; most of the changes were reflected in the Berlinetta and standard sport coupe models. All the '86 sport coupes came equipped with 15-inch styled wheels and a performance exhaust system. This was also the year that a "center high mounted stop light" (CHMSL") became mandatory. Thus, all Camaros had a third lamp mounted near the edge of the rear hatch.

This was also the year that the Berlinetta was discontinued, with only 4,479 being sold before the model was dropped early in the year. Other options for the Camaro

In 1986, the Tuned Port Injection 305 (5.0 liter) V8 was detuned with a new camshaft, much to the disappointment of performance nuts. The IROC-Zs were still the hot-selling vehicle and the carbureted H.O. engine option was still available.

1986 were rear window louvers, fog lamps (standard on IROC-Zs) and an automatic headlight dimming rearview mirror.

Technical Highlights—The IROC continued to rule the streets in 1986. One subtle change, however, made the '86 IROC a little less desirable—a different camshaft in the LB9 Tuned Port Injection 305 V8. Duration at .050" lift was shortened from 202-deg. intake and 207-deg. exhaust to 179-deg. intake/194-deg. exhaust duration. Valve lift was also reduced from .404" intake/.415" exhaust to .350" intake/.384" exhaust. The alteration in valve timing reduced horsepower from 215 to 190, making it the same as the LB6 High Output, carbureted engine.

All V8 engines were manufactured with a revised crankshaft that was machined to accept a one-piece rear main seal. This seal has proven tremendously more effective at preventing leakage, but the block machining that it necessitates prevents earlier model crankshafts from being installed in 1986-and-later blocks unless a special adapter is installed. The new seal also calls for a specific type of oil pan; pans are not interchangeable between 1985-and-earlier, and 1986-and-later engines.

1LE—Although 1987 marked the official introduction of the 350 cid TPI engine, some '86 IROC-Zs were produced with this engine and also equipped with a five-speed manual transmission. It's not known exactly how many of these IROC-Zs were made, but they were essentially hidden from the public, because the 350 was part of the RPO 1LE package that was developed for road racing in various showroom stock classes, and was not widely advertised. Since the 1LE package was intended for racing, it did not include air conditioning, radio or any power accessories. The 1LE option was available only on standard coupes (without T-tops) and featured specially calibrated springs, shock absorbers and sway bars. For collectors, these cars are definite winners.

1987

This year, which marked Chevrolet's 75th anniversary, was another great one for the Camaro as a number of new exciting options were offered. For the first time since 1969, a convertible was available. However, although a convertible could be ordered as a Regular Production Option, Chevrolet didn't build them. The conversions, which were available on IROC, Z28, LT (the Berlinetta replacement) and base sport coupe models, were performed by Automotive Specialty Company (ASC), an independent subcontractor.

Technical Highlights—The hot ticket beneath the hood of an IROC-Z model was RPO L98, a 350 cid, Tuned Port Injection (TPI) engine that produced 220 horsepower.

Chevrolet called the 1987 IROC-Z a "leaner and meaner machine." With the new 5.7 liter Tuned Port Injected V8, it was one of the hottest selling Chevrolets that year. Convertible models were also made available this year.

The new Camaro LT also came on the scene in 1987. It was a popular model that replaced the Berlinetta, and was also available as a convertible.

It was virtually identical to the Corvette L98 except that it included cast iron, rather than aluminum cylinder heads and standard exhaust manifolds in place of the tubular headers installed on Corvettes. (Over the years, the Camaro version of the 350 TPI engine has been referred to with both L98 and B2L RPO numbers.)

Although the new engine and convertible tops made the Camaro more popular than ever, both of these options were not available together; with a convertible top, much of the chassis's structural rigidity and torsional stiffness needed to handle the extra horsepower and torque of the TPI 350, was eliminated.

From an engineering standpoint, the big news for 1987 was the switch to hydraulic roller camshafts in V8 engines. Although the cam profiles were not changed, the switch to roller lifters was credited with an approximate 3.5% improvement in fuel economy and a five horsepower, five lbs-ft. increase in output. Complementing the new valvetrain was a redesigned cylinder head on LB9 engines (305 TPI) that featured a more centrally located spark plug position and improved valve cover sealing. These heads are easily identified by valve covers which have four retaining bolts located in the center.

Another change made to the LB9 engines was incorporation of a new High Energy Ignition (HEI) ignition system. Rather than the unitized system of previous years, the new HEI featured a separate coil and smaller diameter cap. This system is 46% lighter than the older HEI model which incorporated the coil in the distributor cap.

1988

The 1988 Camaros remained relatively unchanged from the '87 models, except there were fewer models available. The LT and Z28 model designations were discontinued, leaving only the Sport Coupe and IROC-Z. This was the first time since 1967 that a Z28 option was not included in the Camaro line-up, although it would not be gone for long.

Technical Highlights—For the first time, all Camaro engines were equipped with electronic fuel injection for more efficient and precise metering of fuel delivery. This theme was responsible for the new shift indicator light (for manual transmission-equipped cars only), which would light up at a certain rpm level to indicate that the driver should shift in order to maximize fuel economy. This optimum shift point was determined by the engine's computer. In practice, shifting when the light lit may have saved fuel, but it discouraged spirited driving (basically, no winding up to maximum rpm before speed shifting into the next gear), which is in contrast to why most people buy a Camaro.

Although the 305 and 350 TPI engines (options LB9 and L98 respectively) were carried over from 1987, the carbureted 305 (LG4) was

The convertible models continued to be popular in 1988 and a new dash with instrument cluster gauges became standard on all model Camaros.

put out to pasture. Its place was taken by option number LO3, a 305 cid V8 with a throttle body fuel injection. The electronically controlled injector, manufactured by Rochester Products Division, featured twin throttle bores, each with a diameter of 1.693".

Other technical improvements for 1988 included a new camshaft (202/207-deg. duration at .050" lift, .404"/.415" valve lift) for 305 TPI engines equipped with manual transmissions, which resulted in increased power, and a true serpentine belt accessory drive system. (In 1987, a combination of serpentine and V-belts was used.) Trivia freaks will also find it noteworthy that this was the first year for a "T-handle" dipstick.

Of greater significance is the fact that "guided rocker arms" were incorporated in LO3, LB9 and L98 engines. These rockers feature a stamped-in lip on each side of the pad that contacts the valve stem. With this change, the task of rocker arm *yaw* (side-to-side) control is handled by the rockers themselves, rather than guide plates or the pushrod bores in the cylinder heads.

Other changes made to improve reliability included new composition head gaskets, new timing chain cover (manufactured from .016" thicker material) and improved oil pan drain plug mounting.

1989

After seven years of refinement, the 1989's were the best handling, best performing and most powerful of the third-generation Camaros to date. By this time, Camaros had become so popular, it seemed that every enthusiast with gas in his veins and a Chevy heartbeat had to have one. Unfortunately, not everyone was willing to pay for one. By 1989, the Camaro held the dubious honor of being the most frequently stolen car in the U.S. In an attempt to lower the Camaro theft rate, Chevrolet introduced the "Pass Key" theft deterrent system, as used in the Corvette. The Pass Key system includes a specially coded ignition key that "tells" the car's electronic system that the rightful owner is attempting to start the vehicle. If a key without the specially coded module is inserted in the ignition switch, any attempt to start the car will be unsuccessful (power to the starter is interrupted). A built-in time lag causes a two- to three-minute delay before another attempt to start the engine can be made. As a result of using the Pass Key system, Camaro thefts dropped dramatically, as did insurance rates. The Pass Key system has been used on Camaros since this model year.

Technical Highlights—This was also the year that the RS replaced the base Sport Coupe and featured styling similar to the IROC-Z. The standard engine for the RS versions was a 173 cid Multi-Port Fuel Injected (MFI) V6; the 305 cid LO3 V8 with Throttle Body Injection (TBI) was optional. The standard engine for the IROC-Z was the LO3 while the 305 cid LB9 and 350 cid L98, both with TPI, were optional.

One change in the TPI engines was the elimination of the 9th injector. In 1988 and earlier models, a separate cold-start fuel injector sprayed fuel into the intake manifold to provide the enrichment necessary to start a cold engine. Chevrolet engineers were able to eliminate this injector by recalibrating the system so that the additional fuel was supplied through the regular injectors. A switch was also made to Multec injectors, which improved fuel atomization, quicker injector action and lower operating voltage.

While these changes improved driveability and reliability, a new, optional dual catalytic converter exhaust system provided a bump in horsepower for TPI engines. Although a "Y" pipe connected the outlet side of the converters to a single muffler inlet, the reduction in back pressure, compared to a single catalytic converter, was significant enough to deliver an increase of approximately 10 horsepower. With this reduced back pressure, horsepower increased to 235 for the 305 engine and 245 for the 350.

Also new for '89 were optional (on IROC-Zs) 16-inch aluminum wheels and Goodyear Z-rated tires. This tire/wheel combination improved handling, stability and traction tremendously. In fact, some magazines claimed that the larger wheels helped them achieve .9 g in skid pad tests.

1990

By 1990, the Camaro had become a proven winner, especially on the race track, where it won several consecutive championships in the Sports Car Club of America's Escort Endurance Challenge and proved to be competitive in IMSA GTO and Trans-Am.

In the SCCA Escort Endurance Challenge, the IROC-Z Camaros dominated the competition, mainly comprised of Mustangs and Porsches. Chevrolet development engineer John Heinricy and racer Tommy Morrison won many races with their special factory-optioned Camaros. With stock horsepower figures of up to 230 for the 305 V8 with five-speed manual transmission, and the improved handling that came along with the 1LE package, these Camaros were always at the front of the pack.

1LE—Although the 1LE package had been available since 1986, this was the first year that many Camaro enthusiasts found out how to order it. Consequently, quite a few of these special order, high performance Camaros found their way onto the streets.

The 1LE Camaros were delivered with front and rear disc brakes, a limited slip differential with 3.42:1 gears, an aluminum driveshaft, larger

The RS model made its debut in 1989 and featured a peppy V6 as its base engine. This "entry level" Camaro may have lacked some of the muscle and cornering power of its big brother, the IROC-Z, but it had just as much style.

Camaros have always made excellent race cars. They've compiled an enviable record in both drag racing and road racing competition. Photo by Michael Lutfy.

12-inch front disc brakes with aluminum calipers, an engine oil cooler and fuel tank baffles. Also available was a choice between seven different spring rates and slightly larger front and rear anti-roll bars.

This package could only be ordered without air conditioning and front fog lamps, although these items could have been added at the dealership. You couldn't just fall off a turnip truck and order a 1LE Camaro; each dealership had to contact Chevrolet for specific ordering instructions. Most racing version 1LE Camaros were ordered with the 5.0 Liter, five-speed manual transmission. According to Chevrolet engineers, several 350 V8 1LE versions were built, although an exact figure is not known. Consequently, a 1990 1LE Camaro with a 350 TPI L98 engine is an exceptionally rare car with potential value. Collectors take note.

Technical Highlights—For 1990, a major change in both the 305 (LB9) and 350 (B2L) TPI engines was a switch in the fuel control method, from mass air to speed density. With a mass air system, a sensor (known, obviously, as a mass air sensor) measures the amount of air entering the engine and adjusts fuel flow accordingly. A speed density system relies on inputs from manifold pressure, manifold air temperature, crankshaft reference and throttle position sensors to calculate the amount of incoming air and adjust fuel flow accordingly. Elimination of the mass air sensor improves reliability (one less component) and does away with a potential flow restriction, but speed density systems don't offer the flexibility of their mass air counterparts.

As far as safety was concerned, this was the first year for a supplemental inflatable restraint system, or more commonly known as an air bag. RS Camaros also received a bigger 191 cid (3.1-liter) MFI V6, which proved to be a winner with prospective Camaro owners who couldn't afford the higher priced IROC-Z. RS owners could have opted for a 305 cid V8 by ordering RPO L03.

1991

After eight years with the same body style, the engineers at Chevrolet decided to make some changes. Although the basic design of the '91's remained unchanged, a new aero package was implemented which included openings in the lower rocker panels and the addition of a rear wing on the performance models. This was also the year that Chevrolet did not renew their contract with the International Race of Champions, so the Camaro was no longer the vehicle used for this racing series. Thus, the term IROC was no longer associated with the Camaro. The performance models went back to using the Z28 badge and Chevrolet made no fuss (meaning no formal announcement) about dropping the IROC name.

The new Z28 Camaro was similar to the previous year's version except for a minor face lift that included slightly different hood ornamentation and a revised front fascia. New 16-inch wheels, graced with new P235/55R-16 tires (as opposed to the earlier P245/50R-16 size) added better looks and handling. The 16-inch tire/wheel combination was an option on RS models and Z28 convertibles but standard on Z28 coupes. Speed-rated P245/50ZR16 Goodyear radials were also optionally available on the Z28 coupe and standard on the convertible.

The interior of all Camaro models included brighter yellow graphics for the instrument clusters and controls and a few new exterior colors, although some of them were not made available to the Z28 versions.

By 1991, even the police had realized the Camaro's appeal so Chevrolet offered a "Special Service" package for the RS model. Designed for "highway traffic law enforcement," option B4C included dual catalytic converters, 16" aluminum wheels, four-wheel disc brakes, engine oil cooler, 105-amp alternator, 145 mph speedometer, air conditioning and performance suspension. Either the 305 LB9 with five-speed manual or the 350 L98 with automatic could be ordered.

1992

Although the Camaro turned 25 this year, Chevrolet did little to commemorate it. Although there were rumors of a special Camaro with a killer engine (the LT-5 from the Corvette ZR-1, for example) about the only distinction was a 25th Anniversary badge mounted on the dash. The styling was exactly the same as the 1991 versions except that Chevrolet decided to bring back the rally stripes that extended from the hood to the rear hatch as an option. This option was the Heritage

In 1991, the Camaro lost its IROC-Z designation as Chevrolet opted not to participate in the racing series. With Dodge Daytonas becoming the vehicle used in the racing series, Chevrolet returned to the legendary "Z28" moniker for the top-of-the-line Camaro.

The year 1992 marked the 25th Anniversary of the Camaro and a Heritage Appearance Package was made available in both Z28 and RS models.

Appearance Package and was available only with Z28 and RS models in Arctic White, Bright Red or Black. Some of the other new hot colors for 1992 included Purple Haze Metallic, Dark Green Gray Metallic, Polo Green II Metallic and Medium Quasar Blue Metallic. All of these colors really made the '92 Camaros stand out in a crowd.

Technical Highlights–The Z28 models also acquired additional ability

Camaros seem to beg for performance-improving modifications. Even Chevrolet engineers have heard the call. This factory "project car" features a 454 cid big block and electronic fuel injection. File this under "things we'd like to see."

to stand out going around a corner. A new optional performance handling package included larger 36mm tubular steel front and 23mm solid steel rear anti-roll bars and stiffer springs which improved handling considerably. The standard Z28 anti-roll bars measured 34mm on the front and 21mm on the rear. RS anti-roll bars were 30mm in the front and 18mm in the rear with the base tires but were the same as the Z28 when 16-inch tires were ordered.

The drivetrain remained the same for the Z28 with the 5.7 liter V8 heading the pack, followed by the 5.0 V8. RS models again were available with the 3.1 (191 cid) V6 or the 5.0 V8 with Throttle Body Fuel Injection.

Interior highlights for '92 included three-point safety belts for rear seat passengers on convertible models. Along with more usable trunk space

Another factory project—an LT-5 engine from a ZR1 Corvette shoe-horned into a Camaro engine compartment. Put this in the same file as the 454 Camaro.

Chevrolet also experimented by installing a carbureted 454 in a third-generation Camaro. It's a great swap, as long as the spark plugs don't need changing.

on convertibles, a new assist mechanism reduced the amount of effort needed to raise and lower the top.

Although there have been many changes throughout the history of the Camaro, it will always be one of America's best performance values and serve as a solid base for building a street machine. The '82 to '92 Camaros also offer attractive, aerodynamic styling which provides them with visual appeal that's perfectly matched to their performance potential.

With the introduction of the fourth-generation Camaro in 1993, a new chapter was begun in Camaro history. But like the 1955 Chevy, third-generation Camaros will continue to grow in popularity among car enthusiasts for decades to come.

All late-model Camaros employ a computerized engine management system which is essential for combining low exhaust emissions with strong performance. Thanks to revised PROM calibrations, this Camaro's performance matches its aggressive appearance.

COMPUTERIZED ENGINE CONTROL SYSTEMS

When the third-generation Camaro debuted in 1982, computerized engine control systems, also known as *electronic engine management systems*, were rapidly becoming the norm on all American production vehicles. Although these systems may be intimidating at first glance, (although fools do rush in where angels fear to tread) they are merely an electronic means of accomplishing what was formerly a mechanical function—the control of ignition timing and fuel delivery.

The computers (formally called an Electronic Control Module or ECM) used for engine management rely on a group of sensors to supply operating condition data. This information is then processed as a means of determining the appropriate response, which takes the form of electronic signals that serve to advance or retard ignition timing and increase or decrease fuel flow, as required by changes in operating conditions. These are the same functions that are controlled by carburetor jets, power enrichment circuit and accelerator pump in a conventional carburetor, and by centrifugal and vacuum advance mechanisms in a mechanical distributor.

SYSTEM COMPONENTS

However, computers offer much more precise control because they can respond much more quickly than mechanical devices and monitor a greater number of variables. The following is an explanation of the devices employed in a typical onboard computer system (such as the one used on TPI engines).

Electronic Control Module (ECM)—The brains of the outfit, the ECM receives all sensor input, performs the calculations required to determine fuel flow rates and ignition timing and sends out signals to put the results of those calculations into effect. ECMs are also programmed to detect sensor failures and abnormal operating conditions; a trouble code is set whenever an abnormal operating condition is detected.

PROMs contain air/fuel and ignition calibrations that are specific to each year and model of vehicle. "Power Chips" such as those produced by Hypertech, contain revised programming that's designed to improve power at wide open throttle. Hypertech chips are CARB (California Air Resources Board) exempt, so they're legal in all 50 states.

PROM—The letters stand for Programmable Read Only Memory (which means that once programmed, a computer can read it, but can't alter it). The PROM, which is also called a *chip*, holds all the data that the ECM needs to match a given set of sensor inputs to fuel and ignition control outputs. The data inside a PROM is arranged in a multi-dimensional array that looks like a topographical map. The reason that computer-controlled engines don't always respond well to radical modifications is that if the ECM can't match the sensor input to the data in the map, it can't come up with the proper output data. The computer continues to "hunt" for a recognizable combination of input data, and does the best job it can in coming up with proper output data—which makes for erratic engine operation. However, aftermarket PROMs can be programmed to include conditions not recognized by a standard PROM. So all it takes is the right programming to allow a General Motors ECM to handle virtually any engine—regardless of the extent of modifications. However, extensively modified engines, and the PROMs needed to control them, will probably not be emissions-legal.

Exhaust Gas Oxygen (EGO) Sensor—Also called an O_2 Sensor, it measures the amount of oxygen in the exhaust and alters fuel flow so that the air/fuel ratio is maintained at *stoichiometric*, which is the chemically ideal air/fuel ratio of 14.7:1. One of the problems with standard EGO sensors is that they must be located relatively close to the cylinder heads so they reach operating temperature. Heated EGO sensors are a solution used on many original equipment and custom installations where the sensor can't be located close enough to the heads to assure proper operation. A heated EGO sensor may be mandatory when an engine is equipped with headers because the sensor location is frequently too far downstream and the exhaust is too cool for proper operation.

Manifold Air Temperature Sensor (MAT)—As its name implies, this sensor keeps track of the air temperature within the intake manifold. This data is used by the ECM so that calibrations can be fine-tuned to accommodate changes in the temperature of incoming air.

Throttle Position Sensor (TPS)—It doesn't take a rocket scientist to figure out that the TPS tells the ECM the position of the throttle. What isn't so apparent is that this sensor also sends data concerning the rate of throttle opening. This data is used to calculate the degree of air/fuel enrichment required for acceleration (in a carburetor this function is handled by the accelerator pump). The TPS position is physically adjustable, and a digital volt meter can be used to determine the proper position. When placed in the stock position, the TPS will have an output of between .53 and .55 volts with

Mass Air Flow (MAF) sensors are included as part of the induction system on all 1985-1989 Camaro TPI engines. The MAF sensor measures the amount of air entering the engine and transmits this information to the ECM. In turn, the ECM plugs airflow data into the equation that it performs to come up with the amount of fuel required to achieve the proper mixture under all operating conditions. Photo by Dave Emanuel.

the throttle closed. Rotating the TPS counterclockwise increases the voltage which results in a richer air/fuel ratio, but if the sensor is moved too far, the "Check Engine" light will illuminate during idle. This condition can usually be avoided by keeping TPS voltage below .64 with the throttle closed; .54 is the recommended setting in almost all instances.

Coolant Temperature Sensor—The name is self-explanatory and the output of this sensor has a pronounced effect on performance. Both air/fuel ratio and ignition timing are altered depending upon coolant temperature. Lower temperatures result in richer mixtures and more aggressive spark timing. High temperatures have the opposite effect.

Knock Sensor—Typically located in a coolant drain hole on the lower side of the block, a knock sensor tries to protect the engine from itself. The sensor itself is like a small microphone and when it "hears" detonation, it sends a distress message to the ECM, which responds by temporarily retarding timing. The ECM returns timing to its original setting when detonation is no longer detected.

Idle Air Control (IAC)—This device incorporates a small electric motor that is used to alter idle speed according to changing operating conditions. The motor opens or closes a valve to admit more air or less air as required to maintain the desired idle speed.

Manifold Absolute Pressure Sensor (MAP)—Manifold absolute pressure is another way of stating manifold vacuum. The MAP sensor is essentially an electronic vacuum gauge that tells the ECM the amount of load under which an engine is operating. The ECM alters air/fuel ratio to accommodate varying load conditions, just as a power valve or metering rods perform this function in a carburetor. This sensor is used with throttle body and TPI speed density systems, but not with mass air flow systems.

Mass Air Flow (MAF) Sensor—Found on 1985-1989 TPI systems, the MAF sensor measures the amount of air flowing into the intake manifold. The ECM calculates airflow by monitoring the voltage drop across a thin heated wire that's positioned in the air stream. As airflow increases, it cools the wire, so the voltage drop decreases; at wide open throttle, voltage through the wire (not voltage drop) will be close to the 5-volt maximum, while at idle, voltage will be about .4 volts. An MAF sensor allows a control system to accommodate a wider range of conditions because it actually measures the amount of air flowing into the engine (By comparison, speed density systems calculate airflow.) However, in an effort to reduce production costs, Chevrolet eliminated MAF sensors on TPI engines beginning in the 1990 model year.

Limp Home Mode—Of course, with all of these sensors, something can very well take a wrong turn, so electronic engine control systems incorporate what's known as a "Limp Home" mode. If there is a major system or sensor malfunction,

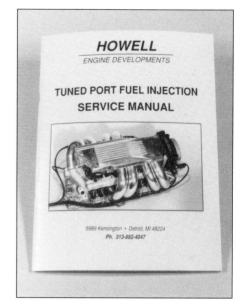

If you plan to "play" very much with a TPI engine, you should have a copy of Howell Engine Developments' "Tuned Port Fuel Injection Service Manual." It contains a complete list of trouble codes and service procedures and some details of the inner workings of GM computer systems.

Laptop computers and hand-held control pads are replacing vacuum gauges and dwell meters as diagnostic tools. This is true for both stock computers and custom aftermarket models.

the ECM automatically slides into "Limp Home" mode whereupon it holds timing to about 22 degrees and establishes a relatively rich air/fuel mixture. The whole idea of the "Limp Home" mode is to protect the engine from damage while it is being driven a short distance. Some custom wiring harnesses (designed to allow installation of TPI systems in vehicles not originally so equipped) force activation of "Limp Home" mode. This isn't desirable, nor is it necessary. Proper harnesses, such as those manufactured by Howell Engine Developments (Clinton Township, MI), are available from a variety of sources.

Trouble Codes

The ECM responds to extreme sensor inputs by setting a trouble code and illuminating the "Check Engine" or "Service Engine Soon" indicator. Trouble codes can be read by connecting a "scan tool" to the Assembly Line Diagnostic Link (ALDL) connector located under the dash on the driver's side. They can also be read by attaching a jumper between terminals "A" and "B" of the ALDL.

Before attaching a jumper wire, the ignition should be turned on (but the engine should NOT be started)

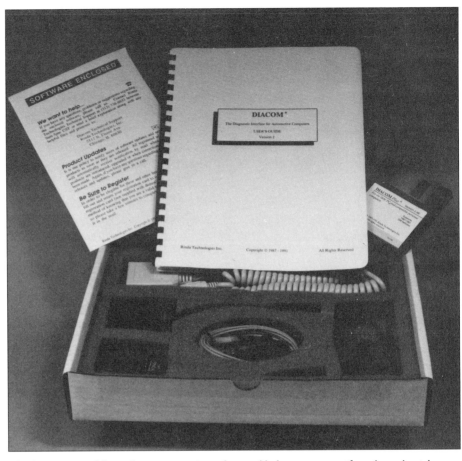

The diagnostic capabilities of a Camaro's ECM makes troubleshooting easy—if you have the right equipment. The Diacom system from Rinda Technologies is the ideal tool for "dumping" the ECM. In addition to locating problems, Diacom can be used to develop custom PROM calibrations when necessary. Photo by Dave Emanuel.

and the "Service Engine Soon" or "Check Engine" light should remain illuminated. When the jumper is attached to terminals "A" and "B," the light should flash a "12" code—

An underdrive crankshaft pulley (left) is noticeably smaller than its stock counterpart (right) and therefore reduces the speed at which accessories are driven. In turn, parasitic horsepower loss is minimized so more power is available at the rear wheels. Operation of the ECM is unaffected.

one flash, a short pause, then two flashes. After another pause (noticeably longer than the one between the single flash and double flash) the "12" code will flash two more times. After that, any trouble codes stored in the ECM's memory will flash three times, in numerical order. After all stored trouble codes have been read, a "12" code will flash again.

Trouble codes can be set by either intermittent or continuous conditions. If, for instance, a loose connection at one of the sensors triggered a trouble code, the "Check Engine" light would illuminate as long as the connector was not making contact. But if it did make contact for more than 10 seconds, the "Check Engine" light would go out, however, the trouble code would

Performance and clean air can live together happily. While retaining the original 305 TPI engine, this car ran 14-second ETs while remaining completely emissions-legal. Photo by Dale Wilson.

remain in memory until power to the ECM was interrupted. Similarly, a hard failure, such as a sensor that has "died," will set a trouble code and cause the "Check Engine" light to remain illuminated until the offending sensor is replaced, and the ECM's memory is cleared (by disconnecting power).

Most trouble codes are the same for all models of third-generation Camaros, but there is some variation depending upon the system configuration. As an example, on a carbureted 1983 Camaro, a "23" code indicates a problem with the mixture control solenoid circuit. However, on the TPI-equipped Camaros, a "23" code means that the manifold air temperature circuit is malfunctioning. For this reason, the appropriate factory service manual should be consulted for a list of trouble codes.

After the problem that caused the setting of a trouble code is fixed, the ECM's memory should be "dumped." This is accomplished by disconnecting power (by unplugging the ECM, disconnecting the battery or removing the continuous battery fuse for approximately 30 seconds) from the ECM. When power to the ECM is interrupted, all stored trouble codes are erased from memory.

Although a jumper wire is helpful, diagnosing problems is considerably easier with a program known as Diacom (available from Rinda Technologies, Chicago, IL). Diacom runs on any IBM PC-compatible desktop or laptop computer and is furnished with the cables needed to connect the computer to the vehicle's ALDL. Use of a laptop is preferable because it allows monitoring of ECM functions while the car is being driven.

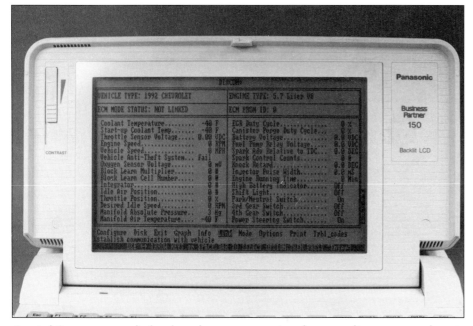

A typical Diacom screen as displayed on a laptop computer. Just about everything you want to know is displayed and up to 18 minutes of data can be stored for future review. Photo by Dave Emanuel.

One of the "speed secrets" associated with increasing the power output of TPI engines equipped with a mass air sensor is to remove the screens found on both sides. When this is done, use of a proper air filter is essential because there's nothing left to protect the sensor's internal fine wire that measures airflow.

Another feature of Diacom software is its storage capability; the standard Diacom stores up to 18 seconds of ECM data, the Diacom-Plus stores up to 18 minutes. This allows a review of conditions after they've occurred, which is helpful in tracking down transient problems. Diacom-Plus also has provisions for graphing up to three functions at a time.

Diacom programs are also used extensively for custom-calibrating PROMS. Many PROM specialists have their customers send them disks containing Diacom data files. The PROM specialist reviews these files, determines where calibration changes are required, "burns" a new PROM and sends it to the customer.

Computers & Carburetors

Computerized engine control systems are not limited to those Camaros with electronic fuel injection. All third-generation Camaros with carburetors (1982-86) also rely on a computer to control air/fuel ratios and ignition timing. Many Camaro owners have been led to believe that they should discard the factory computer system and take a step backwards by relying on mechanically controlled carburetors

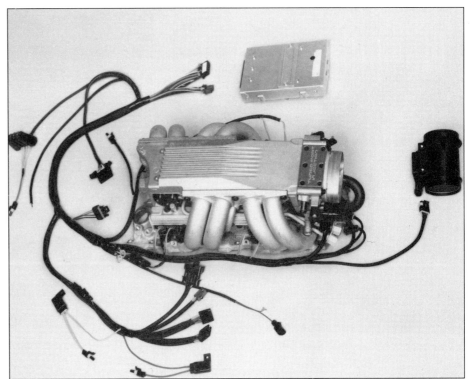

Installation of TPI systems on vehicles not originally so equipped has become very popular. In all cases, a special harness is required. Companies like Howell Engine Developments specialize in building harnesses for all types of custom installations—including retrofit of TPI systems in Camaros originally equipped with a carburetor.

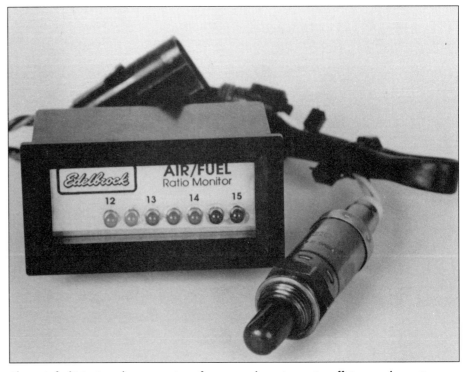

Electronic fuel injection relies on a variety of sensors to determine engine efficiency and operations. One of these is an oxygen sensor that determines air/fuel ratio based on the amount of oxygen in the exhaust. In addition to supplying input to a computer, these sensors can also be incorporated into stand-alone systems that inform the driver about the current status of air/fuel ratio. Such a system is used in addition to the oxygen sensing hardware incorporated with the computer.

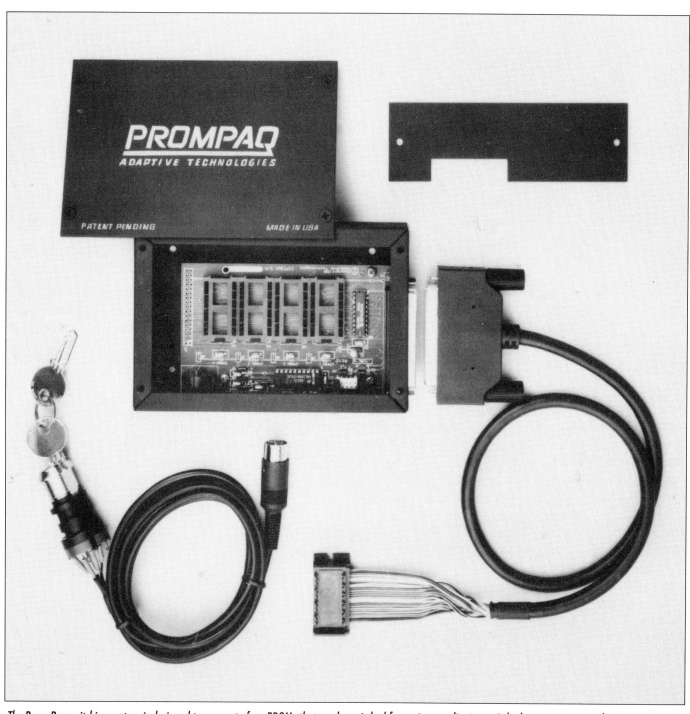

The Prom Paq switching system is designed to use up to four PROMs that can be switched for various applications. A dual-purpose street and strip car can use the stock chip for normal driving and switch to a performance chip for racing situations.

and distributors. This is not only ill-advised, it's also illegal. Federal EPA regulations impose a $2,500 fine to disconnect or render inoperative any emissions-control component. Besides, you're really better off sticking with an electronic system. It will deliver better performance and economy than a mechanical system and will also prevent you from running afoul of the law.

MODIFICATIONS

For increased performance purposes, the only part of the computer that needs to be addressed is the PROM or chip. The PROM for each type of engine contains different calibrations that are specific to that engine/transmission combination. For example, a chip from a Camaro with a 305 cid engine with manual transmission is totally different from the chip for the same vehicle with an automatic.

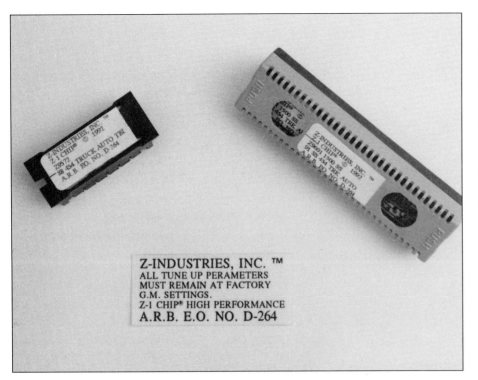

Turbo City in Orange, California, offers smog-legal performance chips for the Camaro. They basically recurve the timing, richen the fuel mixture and delay lockup of the torque converter on automatic transmission-equipped cars.

Irrespective of the PROM or vehicle in which it's installed, GM ECMs have a learning capability. During the course of normal engine operation, the ECM reviews the relationship between sensor inputs and fuel control outputs. If it "sees" a trend developing, it modifies its output as required.

As an example, if the air cleaner becomes clogged (which restricts airflow), the ECM will detect that standard fuel calibrations are causing a rich mixture. Consequently, it will reduce injector pulse width as a means of leaning out the mixture. When the air cleaner is changed, the ECM will still be using the new fuel calibrations it has "learned," but the mixture will now be lean. In time, the ECM will "learn" that more fuel flow is required and will make appropriate adjustments. This process can be sped up by "dumping" the ECM's memory. When this is done, the ECM doesn't have to "unlearn" its new calibrations; it is simply forced back to ground zero (when power is disconnected) and starts "learning" again.

The ECM can't make dramatic changes in calibrations, but its learning ability must be considered whenever changes are made. Consequently, it's a good idea to "dump" the ECM's memory prior to starting the engine after there's been a significant change in equipment or operating conditions.

Aftermarket PROMs

There are several manufacturers of aftermarket performance chips, but not all of these chips are created equal. Some are nothing more than copies of GM service chips. However, reputable companies like Z Industries and Hypertech have developed special programming that is significantly different from that of stock PROMs. Many chips have also been granted an exemption by the California Air Resources Board, so they're emissions-legal in all 50 states.

Performance chips improve power the same way that a "power tune-up" increases the output of pre-computer engines. Specifically, they employ more aggressive ignition timing and alter fuel delivery as required to take full advantage of the timing changes. In many instances, higher octane fuel is required to prevent detonation with the more aggressive timing. Compared to a

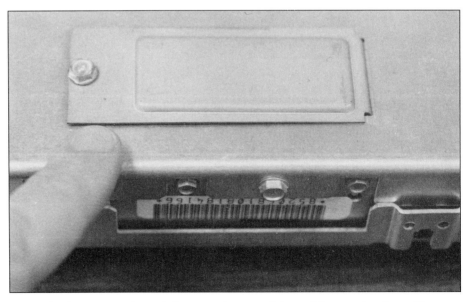

Looking for the PROM? It's beneath this cover. The size and shape of the PROM has changed from year to year and so has the panel used to cover it. However, the cover panels are all similar in appearance and held in place by either one or two screws. Photo by Dave Emanuel.

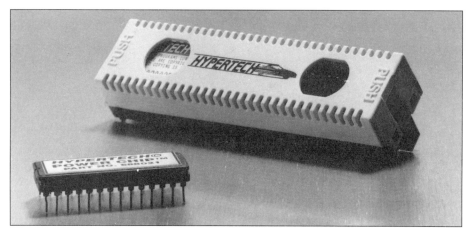

The Hypertech PROM in the background is the size, shape and design of stock PROMS found in most 1987-and-later Camaros. It may be removed and replaced with little chance of damage. The chip in the foreground is the type found in earlier computers. Care must be taken to avoid damaging the pins which fit into the receptacle in the ECM.

stock chip, a high performance model trades-off some ability to run on low octane fuels for increased power.

But be advised that all aftermarket chips, except those that are custom calibrated for modified engines, alter ignition timing and fuel flow only when an engine is running at or near wide-open throttle; in order to meet regulations, aftermarket chips must retain stock calibrations for idle and part throttle operation. So while these chips do offer a definite performance improvement, they are not designed to accommodate a highly modified engine.

Unless a vehicle is used for racing or is otherwise exempt from emissions regulations, chips with custom calibrations without CARB certification or any other form of legal exemption, are considered illegal by the clean air police. However, there are a number of companies that are involved in road racing, so if your Camaro is a legitimate race car, custom-calibrated chips are available.

Replacing a PROM

If a chip change is in order for your particular project, installing a performance PROM is not as intimidating as it may seem (see sidebar nearby). Before beginning this procedure, make sure the ignition has been turned off. Then locate the computer, which in 1982-92 Camaros is hidden behind the dash on the passenger's side. Several small screws secure a trim panel to the underside of the dash. Once this panel is removed the computer is accessible. On '82 to '90 model Camaros, the computer is held in place by a plastic bracket that's attached to the dash with two screws. When these screws are removed, the computer may be pulled down and out from behind the dash. 1991 and 1992 models use a large piece of Velcro to hold the computer to the bracket and thus it can just be pulled down.

Once the computer is out in the open, you'll notice a small cover plate that's held in position by two screws. Unscrewing these exposes the chip which can be easily removed by pulling away the two locking tabs and gently lifting the chip up out of its socket. An aftermarket chip can then be plugged in and the computer can be installed.

Any chip should fit into place with a gentle push. If extreme pressure is required to seat a chip, chances are it's not positioned properly or is not designed for that particular socket. Chips are "keyed," usually with a small tab, so they cannot be inserted incorrectly. A chip should never be forced into place. If it will not seat properly in its socket, remove it and make sure its tab is positioned to fit into the slot in the socket.

Before starting the engine, it's a good idea to disconnect the battery for about 30 seconds. This will clear any trouble codes that may have been stored and will also bring the ECM back to base zero, by eliminating any "learned" data.

Special Computers & Conversions

With a custom calibrated PROM, the GM computer can handle virtually any modified engine—even ones that aren't intended for operation on the street. So an aftermarket computer is required only if the existing ECM is not compatible with the ignition and fuel system installed on an engine. Aftermarket computers, which are designed to be programmed by the user, may also be required if a company with custom

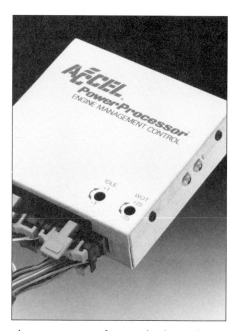

The Power Processor from Accel is designed to allow easy changes in calibrations. This computer is engineered for use with speed density systems and is not emissions-legal.

23

INSTALLING A POWER CHIP

When a Camaro engine's air/fuel and ignition calibrations are controlled by a computer, the settings, relative to a particular engine/chassis combination, are contained in a removable computer chip called a Programmable Read Only Memory (PROM). This is the same system used on all GM-built vehicles. Consequently, if a quicker advance rate, or a richer or leaner air/fuel ratio is desired, it is most easily achieved by changing PROMs. Prior to the days of computerized engine controls, these modifications were accomplished by changing carburetor jets and recurving the centrifugal advance.

There's no magic to be enjoyed through a change in PROMs (although sometimes there appears to be). Proper calibrations simply allow an ECM to respond properly to engine conditions. The best part about a PROM change is that it can be accomplished with a minimum of effort, yet results can be dramatic.

Note that as a vehicle is driven, the ECM "learns" to make changes in the air/fuel calibration (if necessary). It's common practice to "dump" the computer's memory whenever a change is made. However, this isn't always necessary or desirable. If clogged injectors or inadequate fuel pressure are causing an engine to run lean, that condition will still exist after a PROM change. Erasing the computer's memory simply puts the learning function back at square one. Ultimately, it must readjust the fuel calibration in the same direction that it had moved previously. In this case, it would make more sense to maintain the information that has been learned and allow the computer to make changes from that point, rather than forcing it back to square one.

In any instance, any time a PROM change is made, or the computer's memory dumped, the car should be driven for a few miles before any evaluations are made. This will allow the computer to make any fuel calibration adjustments that might be required for optimum performance.

1. The computer is located behind the dash on the passenger's side. By removing a few screws, the cover panel beneath the dash can be removed, exposing the computer.

2. The computer is held in place by a bracket that is attached with either screws or velcro (depending on model year). Use the appropriate tools to loosen the bracket and pull the computer down, being careful not to kink any wires.

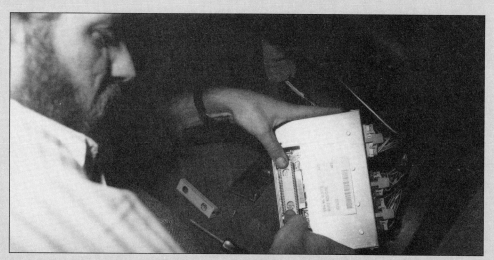

3. One or two bolts hold the PROM access panel in place, and once removed, the PROM is accessible. PROM configuration varies according to model year, but all are removed by gently lifting or prying. Large PROMs used in most fuel-injected Camaros have locking tabs at each end. When these tabs are pushed back, they lift the PROM up so it can be easily removed. To install a new PROM, simply reverse the procedure.

4. As a vehicle is driven, an ECM "learns" if the preprogrammed calibrations are accurate and makes minor adjustments in the fuel curve if required. If desired, the data that has been "learned" during previous driving cycles can be erased by interrupting power to the ECM. This is most easily accomplished by pulling the ECM power fuse that's located beneath the hood, on the passenger's side.

A special carrier (the black plastic piece that surrounds the chip in this photo) is used with exposed pin type chips to make installation and removal easier. Even so, the changing operation must be done carefully to avoid bending one of the pins. Photo by Dave Emanuel.

PROM calibrating capabilities cannot be located, or if frequent changes to calibrations are anticipated (as would be the case with a race car).

Add-On Devices—Additionally, add-on and "piggyback" devices, which plug in to the GM ECM and modify its output signals, are available. One such device, the "Smart Spark" manufactured by Digicon Engineering (North Vancouver, Canada) allows base timing to be increased or decreased up to 19 degrees and also includes a spark advance map which can be programmed to alter spark timing every 500 rpm up to 6000 rpm. This arrangement allows the car owner to optimize spark timing by simply dialing in a specific number of degrees that are either added to or subtracted from the stock spark curve.

Aftermarket Computers—The computer business is somewhat volatile, so it's not unusual for companies to come and go quickly. However, Motec Systems (Huntington Beach, CA), Accel (Branford, CT) and EFI Technologies have been around for some time, so it's reasonable to expect that these companies will be around to support their products in the future.

The Motec system may be programmed through either a hand-held controller or IBM-compatible personal computer and typifies the high level of sophistication that's available with aftermarket computerized engine controls. The standard system controls both the fuel and ignition systems, but a lower cost model, with only fuel management capabilities is also available. Motec also offers a selection of fuel injectors, fuel pumps, pressure regulators and sensors which are required to create custom fuel and ignition management systems.

TPI engines are equipped with eight injector nozzles that are controlled by the ECM. Fuel flow is controlled by pulse width—the longer the injector on time, the greater the amount of fuel flow. Flow rates do change as injectors age and accumulate deposits, so they may need to be replaced or cleaned periodically. Photo by Dave Emanuel.

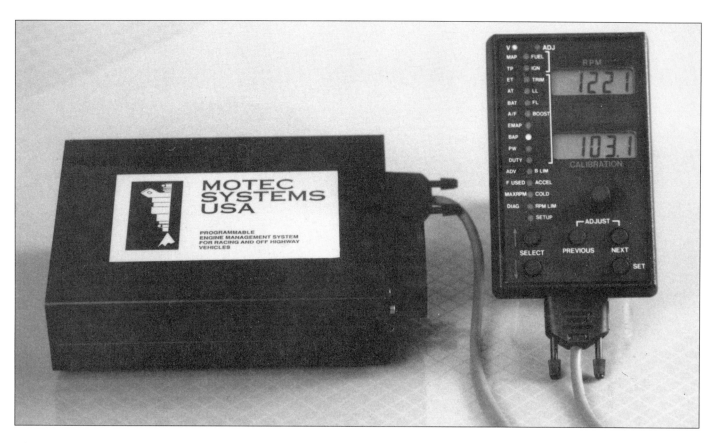

The engines found in full-on road racing Camaros are often equipped with a sophisticated engine management computer designed specifically for racing. As opposed to a stock ECM, which relies on a preprogrammed PROM for calibrations, race computers, like this one from Motec Systems, can be easily programmed with either a laptop computer or hand-held controller. This allows changes in calibrations to be made quickly.

Although aftermarket systems offer tremendous versatility, and are the only viable option for many race applications, they require extensive knowledge on the part of the user. With no basic calibration from which to begin, the user must have the ability to develop a complete map to cover all operating conditions. Mapping usually is begun with the engine on a dyno and then fine-tuned after installation in the vehicle. Although these systems are used on street-driven vehicles, their use is limited because of emissions system requirements. So even though an aftermarket fuel injection system can produce cleaner exhaust than a stock carbureted system, it's illegal unless the manufacturer has obtained EPA acceptance or a California Air Resources Board exemption.

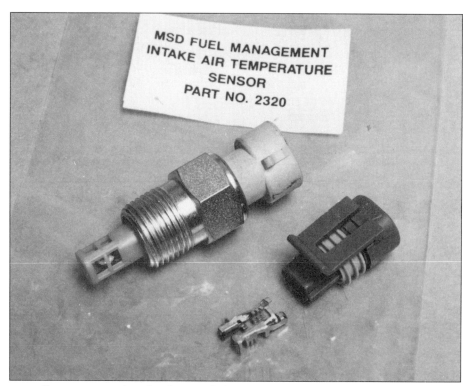

MSD offers an extensive line of replacement fuel injection parts. In addition to the injectors themselves, the company also offers a variety of sensors for TPI and throttle body systems. Photo by Dave Emanuel.

Late-model Camaro engines are especially sensitive to ignition timing. Fortunately, the stock factory unit is very good.

IGNITION SYSTEMS 3

Late-model engines are especially sensitive to ignition timing. Besides the initial distributor timing, there are other factors that influence ignition timing at any given circumstance. The factory computer system takes these factors—air temperature, rpm, acceleration, engine load, throttle position—into consideration and optimizes timing. The only way to change the spark curve is through altering base timing or changing the PROM.

Fortunately, '82-'92 Camaros are blessed with a very good, factory ignition system. In fact, up to 4500-5000 rpm, it puts out as much, or more spark energy than most aftermarket ignitions. There is not much you can do to increase your Camaro's ignition output. However, proper maintenance is required to ensure that it operates up to its full potential.

To begin with, there are several types of ignitions used in late-model Camaros. On four- and six-cylinder engines, either an electronic, breakerless distributor or a direct fire, computer-controlled coil pack is used ('87-and-later). All V8 models use one of two styles of HEI ignition systems. The first type, used on '82-'86 models, uses the conventional HEI system with the coil located on the distributor cap. Later models use the remote coil HEI system.

Spark Plug Wires

Although the four-cylinder and V6 ignition systems can't be enhanced further than a good set of 8mm spark plug wires, the HEI systems can be optimized a little further. As with the systems for engines of lesser cylinders, V8 ignition systems will benefit from installation of helical wound (spiral core) 8mm plug wires. The spiral core wires suppress RF (radio frequency) noises that can possibly interfere with the operation of computer systems. Some of the better wires available are manufactured by MSD, Mallory, Moroso, Accel and Jacobs Electronics. For late-model Camaros, spark plug wires need to be of the 90-degree boot style on both ends. Be sure to route wires away from the exhaust manifolds, and keep the wires for cylinders 5 and 7 separated to prevent inductive crossfire.

Countless hours have been spent dyno testing computer-controlled engines in an effort to maximize horsepower. Although the computer can make up for many deficiencies, it can't overcome a weak ignition system. Photo by Dave Emanuel.

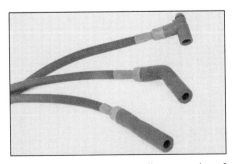

Even a killer ignition system will come up short if the plug wires burn through. Heat-resistant sleeving is often used to protect the wires where they pass close to headers or exhaust manifolds.

Camaro engines equipped with a unitized HEI distributor will benefit from installation of a high voltage coil, such as this HEI Power Coil from Hypertech. It supplies 53,000 volts and allows use of wider plug gaps for increased power and efficiency.

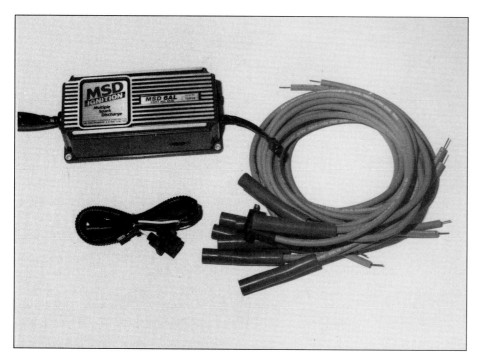

Another means of increasing spark output is to install an MSD 6A or 6AL module. MSD offers a special wiring harness, shown at bottom left, that makes installation a breeze on engines with a separate coil.

AFTERMARKET IGNITION SYSTEMS

The installation of an aftermarket ignition module to enhance stock HEI system performance usually proves to be beneficial. Some ignition boxes enhance the stock ignition by energizing the spark and providing a higher voltage over a wider rpm band than the stock coil can produce. Some systems also provide higher voltage and more than one spark to ensure a proper and more efficient combustion.

MSD

By far, the most effective aftermarket ignition system to install on a late-model Camaro is the MSD-6A unit from Autotronic Controls. The MSD is a multiple spark discharge unit that delivers multiple sparks to each cylinder at relatively low rpm. The MSD units produce a spark that burns for 20 degrees of crankshaft rotation (on an eight-cylinder engine). The electronics generate the maximum number of sparks permitted by the time allowed. But there ultimately comes a point where a second spark can't be generated within the time frame defined by 20 degrees of crankshaft rotation. So maximum energy is put into one long, fat spark when an engine is operating at speeds above approximately 2000 rpm.

MSD units are available with a special harness that tremendously simplifies installation. With one of these harnesses, an MSD module can be installed in a matter of minutes, without any wire cutting or splicing into the existing harness. However, there is occasionally a compatibility problem between the MSD's increased energy output and the ECM. Plug wires with special resistance values may be required to prevent the ECM from being on the receiving end of spurious inputs.

CROSSFIRE

Once the distributor, coil, cap, rotor, spark plug wires and electronics are all properly introduced and working together harmoniously, the next task is to ensure that the spark energy leaving the distributor cap reaches the appropriate spark plug at precisely the correct time. But there are several detours that can be

One problem with 1982-92 Camaros is that there isn't much room under the hood. The MSD box had to be sandwiched between the inner fender and fender brace. Photo by Dave Emanuel.

Many aftermarket distributors are emissions-legal, but hooking them up to the computer is a problem. All third-generation Camaro engines are computer controlled, which rules out the use of conventional distributors unless the car is not driven on the street.

thrown into the path of unwitting bands of sparks. One of the most detrimental is *crossfire*.

The firing of one plug by the spark intended for another can be deadly, especially in a high performance or racing engine. It usually occurs when the high resistance found in a cylinder during the compression cycle causes the current leaving the rotor to jump to an adjacent terminal (where resistance is lower) and fire the plug to which it's connected. Since that is usually the next one in the firing order, the cylinder has already started filling with an intake charge when the misdirected spark occurs. This charge is ignited and combustion takes place while the piston is completing its intake cycle and transitioning to its compression cycle. The result of such extremely early lighting of the air/fuel mixture, which the upward traveling piston is trying to compress, is excess heat and pressure—and sometimes engine failure. A large diameter distributor cap and matching rotor, maintained properly, are the best insurance against crossfire occurring within the cap.

Inductive Crossfire—Another form of crossfire, and one that is rarely understood, is caused by *inductance*. First, you have to know how a coil operates. Basically, current flowing through the primary winding creates a magnetic field that wraps itself around both the primary and secondary windings, inducing a voltage and consequently current (even though the two windings never touch). If you understand this, you're on the road to understanding inductive crossfire. It usually occurs in the adjacent firing cylinders of a V8 engine—numbers 5 and 7 on a Chevrolet—where the two plug wires in question run close together over a relatively long distance. Inductive crossfire is common in wires with solid metal cores.

In effect, the two wires become a transformer much like a coil. As current flows through the "primary" (the first of the two cylinders to fire) it builds a magnetic field that wraps itself around the adjacent wire and induces a voltage in the "secondary." This results in a weak spark reaching the plug in the corresponding cylinder, which will effectively reduce its combustion efficiency, and reduce power output. The cure is to maintain adequate spacing between wires (use non-metallic wire separators) avoid running wires parallel to one another over a long distance and to install helical core plug wires. In a helical or wire-wound spark plug wire, the conductor is spirally or helically wound around a non-conductive core. By winding the conductor in a spiral, the magnetic field is disrupted, eliminating the possibility of inductive crossfire.

The terminals inside a distributor cap have a rough life and are subject to corrosion and deterioration. Routine inspection can often stop a problem before it starts. Both the cap and rotor should be free of carbon tracks and all terminals should be clean. Photo by Dave Emanuel.

Accel's HEI Super Coil steps up the voltage of stock systems significantly. It simply plugs into the cap of a unitized distributor. Photo by Dave Emanuel.

Spark Plugs

There are various types and styles of spark plugs on the market today. Many have uniquely shaped electrodes which are designed to promote a better spark in the combustion chamber. For most street and strip applications, any kind of improvement with a different style of electrode will not make any noticeable difference.

The main emphasis on spark plug selection on a high performance late-model Camaro should be on spark plug durability and correct *heat range*. The only way to determine which plug lasts longer in your Camaro is to try various brands and check them periodically, noting how long they took for the electrodes to become eroded or rounded. However, the correct spark plug heat range and *plug gap*, play a more important role in performance than you may expect.

Heat Range

Spark plugs come in a variety of heat ranges which determine the temperature of the plug tip. Heat range is established by the path that heat must travel through the spark plug body to reach the shell; the longer the path, the hotter the plug. The terms hot and cold are relative—a plug that's too hot for one engine may be considered too cold for another. That's because combustion chamber temperatures vary considerably, depending upon an engine's state of tune and the conditions under which it operates. When heat range is properly matched to requirements, insulator tip temperature will range between 700 and 1500 deg. F. under normal operating conditions.

Recommendations—With a stock engine, it's best to follow the plug manufacturer's recommendation

33

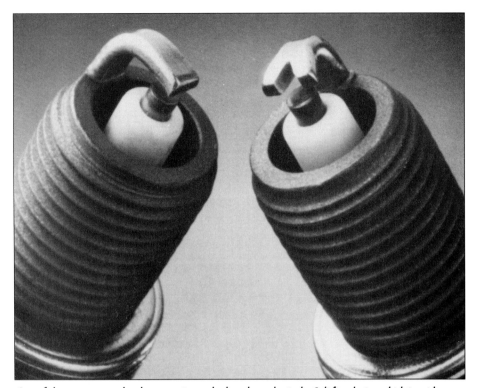

One of the most recent developments in spark plug electrodes is the Splitfire design which is said to fire a broader flame kernel which promotes more efficient combustion. A conventional electrode is shown at left.

If a spark plug's heat range is too cold, it will tend to foul. If it's too hot, this can be the result. Note that the center electrode is completely burned away. Photo by Dave Emanuel.

unless there's a good reason to deviate—such as a modified spark advance curve, poor fuel quality, extreme operating conditions or experience with the recommended plug being inappropriate. With a modified engine, standard recommendations can be used as a basis for determining where in the heat chart to start. Raising compression ratio, leaning the fuel mixture, increasing initial spark advance or reworking the advance curve to come in quicker are all reasons to switch to a plug with a colder heat range.

Cam Duration—Another often overlooked factor when selecting proper heat range is *cam duration*. If no other changes are made, swapping a short duration cam for one with longer duration decreases cylinder pressure, which has the same effect as lowering compression ratio. Shortening cam duration increases cylinder pressure, which has the same effect as raising compression ratio.

So even though a long duration cam is designed to produce more horsepower, at higher engine speeds, under some circumstances, it will create a condition that calls for a hotter, rather than a colder spark plug. On the other hand, if a shorter duration cam is installed as a means of increasing low-speed torque and smoothing the idle, a colder plug may be required because of the increased heat generated by higher cylinder pressure.

Octane—Gasoline quality is yet another factor that influences heat range. Lower octane fuels burn more quickly and create higher combustion chamber temperatures. In years past, when premium leaded fuels were widely available, octane was sufficient to meet the demands of most high-compression street engines. But with the advent of unleaded gas, octane ratings of all grades of fuel dropped. Consequently, preignition, detonation and run-on became more prevalent problems. If you've been running the same heat range plugs for years, but have been plagued by detonation, preignition or run-on for no apparent reason, it just may be that the plugs you're using are too hot because of lower fuel octane. Many times, a switch to colder heat range plugs will eliminate those ugly knock and ping noises.

As you can see, the selection of spark plugs with a proper heat range isn't a cut-and-dried affair. So many factors enter into the equation, that some experimentation is usually necessary to optimize performance and plug life. As a general rule, if you're off on heat range, it's better to be too cold than too hot. Plugs that are too

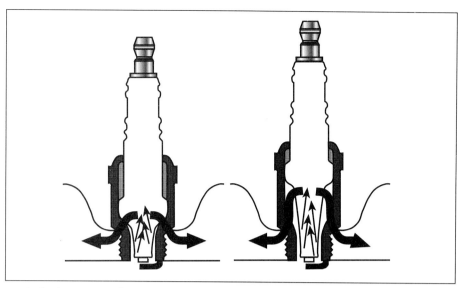
Spark plug heat range is determined by the distance heat must travel from the tip to the shell. The longer the distance, the hotter the plug.

Fresh spark plugs can make a significant difference in both power and fuel economy. Selecting plugs with the correct heat range is also important.

cold will foul more easily, build carbon deposits more quickly and slightly decrease power. Plugs that are too hot can cause preignition, detonation and run-on—all of which can cause your engine to eat its own pistons and valves.

Spark Plug Gap

Spark plug gap also has a definite effect on the long-term wear of the plugs. Normal spark plug gap for late-model HEI-equipped engines is .045". As a general rule, high output ignition systems (such as the GM HEI) can fire a wider plug gap than a standard ignition. Adding an aftermarket ignition module, such as an MSD-6A may allow plug gap to be widened. This is beneficial, because a fatter spark is produced and that leads to more efficient combustion.

Forced Induction—Typically, supercharged, turbocharged or nitrous oxide injected engines will not tolerate as wide a plug gap as naturally aspirated powerplants. The extremely high cylinder pressures

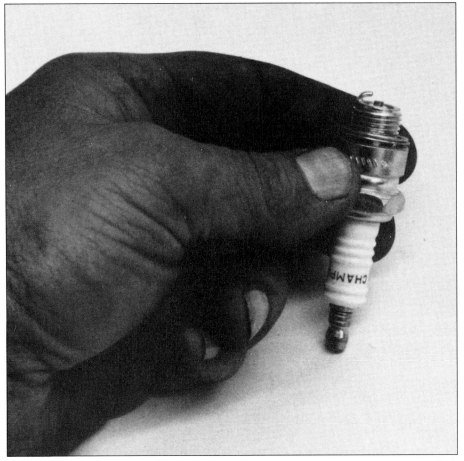

Cut-back ground electrodes are an old drag racing trick for increasing power output. Myron Cottrell of TPI Specialties has documented a consistent five to eight horsepower on a typical 350 small block. Photo by Myron Cottrell

On computer-controlled engines, spark timing can be checked with a conventional timing light, but a quicker and easier method is to connect a Diacom program. Spark advance, shown on the top right of the screen, is one of the variables the program monitors. Photo by Dave Emanuel.

A fully charged battery is essential if maximum ignition system output is to be realized. A good 500+ amp battery should be used. This one has been mounted in the spare tire well through use of a simple battery relocation kit.

associated with mechanical and chemical supercharging create more resistance. The larger the amount of resistance, the more spark energy is required to ionize or jump the plug gap. However, insufficient spark energy is rarely a problem with a high performance ignition system.

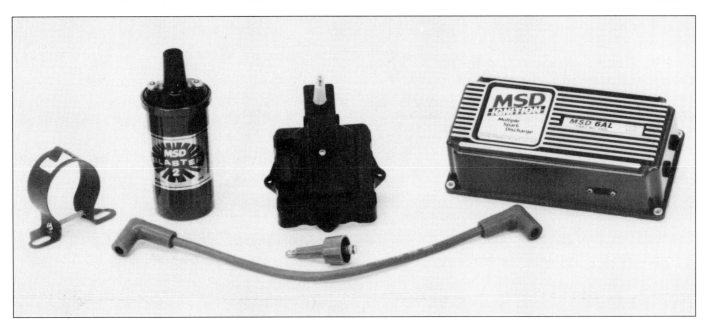

HEI systems with an integral coil can be converted to an external coil system with an adapter kit from MSD.

A low-restriction exhaust system makes improvements you can feel and see at the drag strip. A good "cat-back" system can drop quarter-mile times by one to two-tenths of a second. This project Camaro from All Chevy Magazine runs consistent mid-13 second ETs.

EXHAUST SYSTEMS 4

The addition of a performance exhaust system has always been one of the first items on many enthusiast's list of performance upgrades. Traditionally, this type of modification decreases the amount of back pressure which has always proven to be a relatively inexpensive way to gain more power. Some "experts" have argued however, that to optimize performance, some back pressure is needed. That's undoubtedly true, but with the constraints of the type of exhaust system that is found on 1982-1992 Camaros, it's very difficult to have too little back pressure. Consequently, the name of the game is to minimize it as much as possible.

With a deft hand applied to the exhaust system on late-model Camaros, both V6 and V8 engines will respond with a dramatic jump in Mustang-kicking horsepower.

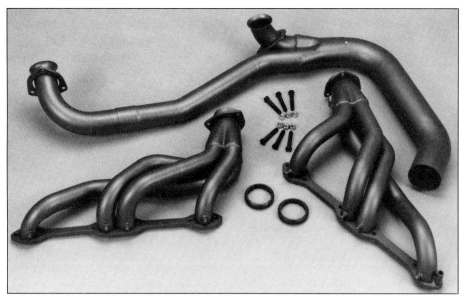

There isn't much room between the engine and chassis in a late-model Camaro, so snaking in a set of headers is no easy feat. Short primary pipes are also the rule because the system must be collected so that it can enter the catalytic converter.

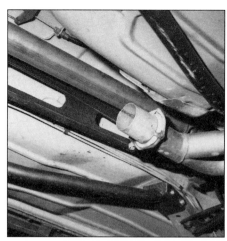

The most common exhaust modification is installation of a cat-back system. This dual catalytic converter system terminates in a 3-inch collector which will be connected to a similarly sized exhaust pipe that will extend all the way back to the muffler.

STOCK EXHAUST SYSTEM DESIGN

Depending on the extent of exhaust system modifications, V8-powered Camaros will realize anywhere from 10 to 30 extra horsepower. But before attempting to improve a third-generation Camaro's exhaust system, it's advisable to become familiar with the stock components. Camaros come with a set of cast-iron exhaust manifolds which connect to exhaust tubing that runs down under the body, which is connected by a Y-pipe on the passenger side. The tubing is made of 18-20 gauge steel and the Y-pipe feeds into either a single or dual catalytic converter system. Dual converters are found only on some 1989-and-later Camaros. From the catalytic converter(s), the factory exhaust is routed straight back to the rear, where it goes over the axle and into a large transversely mounted muffler located just behind the rear axle.

Although the factory exhaust can be modified to reduce back pressure even further, the overall design and layout is very good from a performance standpoint. Because of the way the factory system is routed, it allows for superior ground clearance and does not interfere with the transmission or other drivetrain components that may need to be serviced or modified.

Since it is a good design, it's best to stick with a replacement system that

Looks can be deceiving. This Camaro runs a factory exhaust, complete with dual catalytic converters and a free-flow muffler. The car runs low 13-second quarter-mile times. It's quite a sleeper.

follows the same basic routing. This is especially true with Camaros that have been lowered. Many Camaro owners have made the mistake of attempting to use the old muscle car method of installing a dual exhaust that ends before the axle with a muffler on each side. This type of system falls short on a number of accounts. Not only is it terribly out-of-date, it's also difficult to install, limits ground clearance (even in a Camaro with a bone-stock ride height), and poses a safety hazard. Any exhaust system that terminates in front of the rear axle increases the chance that carbon monoxide will enter the passenger compartment. The best option is to use emissions-legal headers and 3" diameter tubing from the converter back, and follow stock exhaust pipe routing.

CATALYTIC CONVERTERS & THE LAW

Before proceeding any further in discussing exhaust system modifications for late-model Camaros, an examination of emissions regulations is in order. It is a federal offense to alter, remove or render inoperative, any catalytic converter or emissions-control device on any pollution-controlled motor vehicle. Don't let anyone convince you that you can run any kind of test pipe for "off-road use only," or that it is acceptable to hollow out the catalytic converter. The Federal Clean Air Act calls for a $2500 fine for each violation. So if you remove the catalytic converter, pull off the EGR valve and scrap the air injection (smog) pump, you're in for possible fines of $7500.

Cat Back Systems

High performance exhaust systems located to the rear of the catalytic converter are fully acceptable so long as they don't raise noise levels to the point that the local constabulary feels compelled to get up

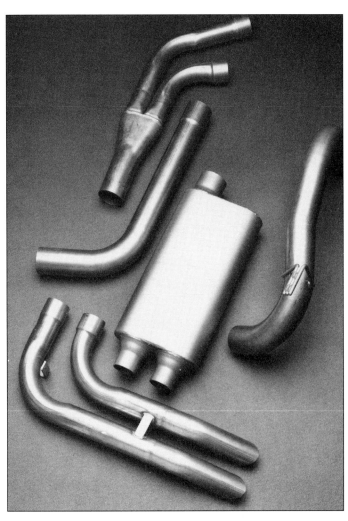

As might be expected, Flowmaster offers complete cat-back Camaro exhaust systems. This one is for use on cars with dual catalytic converters and features the company's high-flow Y-pipe.

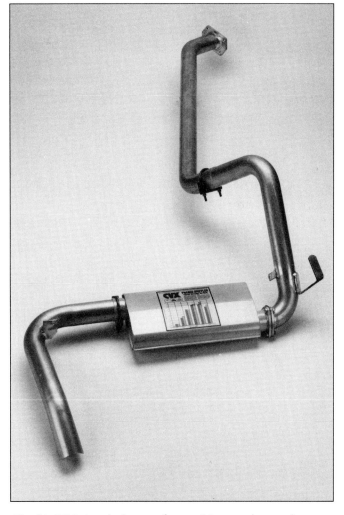

Thrush's CVX 3x3 cat-back system features 16-gauge aluminized mandrel-bent 3-inch tubing and one of Thrush's new CVX phased mufflers with single 3-inch outlet.

Catalytic converters play an important part in keeping exhaust emissions low. They must be maintained for emissions legality but some converters are better than others. This high flow converter is a stock Chevy piece found on most TPI-equipped V8's and is a good replacement for a more restrictive unit.

close and personal with the driver. "Cat back" exhaust systems (which include everything rearward of the catalytic converter), are available from a variety of manufacturers.

Most reputable aftermarket muffler and exhaust system manufacturers create systems that produce an appealing throaty sound that is well within the decibel level range that law enforcement personnel find acceptable. "Cat back" systems are not subject to emissions regulations, since their only function is to control noise, not exhaust gas content. Consequently, noise level is the only concern.

However, keep in mind that all components installed between the engine and the catalytic converter must be EPA accepted or have been granted an exemption by the California Air Resources Board. This mainly applies to headers and the Y-pipe that leads into the catalytic converter(s), as well as the converter itself.

Converter Operation

The catalytic converter is probably one of the most restrictive items in a Camaro's exhaust system. The converter is designed to convert unburned hydrocarbons, carbon monoxide and oxides of nitrogen to less harmful compounds like carbon dioxide and water. The conversion process results from hot exhaust gases coming in contact with a special mesh (that's plated with exotic metals) inside the converter. The metals used on the mesh surface initiate the desired chemical reactions that convert environmentally harmful substances into relatively benign compounds.

Most catalytic converters operate at temperatures ranging from 900 to 1600 deg. F. Although a converter generates heat with the catalytic process, it also requires heat to initiate the chemical action. Consequently, a converter must be mounted as close to the engine as possible. If the exhaust is too cold when it reaches the catalysts, they won't "light off."

Part of the air injection system includes a tube that runs to the catalytic converter. When a converter is changed, this tube must be reinstalled. Photo by Dave Emanuel.

Converter Options

There are several types of catalytic converters that have been used in the Camaro since '82. The least restrictive are the ones used on late-model TPI-equipped engines. Aftermarket companies such as Walker and Thrush market low-restriction catalytic converters that are considerably cheaper than those purchased through a Chevrolet dealer. These converters feature a wide mouth that has a four-bolt flange, making it easy to attach to a 3" exhaust system. Swapping an old-style converter for one of these can make as much as a 5-7 horsepower improvement.

For '89-and-later Camaros, a dual catalytic converter system is used which is probably one of the most free-flowing converter systems on the road today. If your Camaro came equipped with dual converters, you don't need to consider replacing them unless they have ceased to function properly. If they do need to be replaced, they can be ordered from Chevrolet as part number 10185070.

If you can't find a Chevrolet high flow converter for your Camaro, or the price has sent you into cardiac arrest, there are several aftermarket manufacturers such as Accel, Thrush and Dynomax that offer an alternative. However, the aftermarket companies offer a wide variety of catalytic converters, so make sure that the intended replacements have the correct inlet and outlet diameters.

Also keep in mind that you cannot just arbitrarily decide to change catalytic converters. According to Federal law, a vehicle must have logged at least 50,000 miles, and/or the converters must be damaged, missing or non-functioning before they can be replaced. Also note that it is technically not legal to replace a single converter with a dual system if the latter was not originally offered for the specific make, model and year of vehicle. That means it's not legal to install a dual-converter system on a 1988-or-earlier Camaro—unless a complete engine assembly, from a 1989-or-later Camaro is installed.

The heart of any good, emissions-legal system is a good flowing catalytic converter such as this wide mouth model from late-model Camaros and Corvettes.

HIGH FLOW MUFFLERS

Since the "cat back" part of a Camaro's exhaust system is not subject to exhaust emissions regulations, and it is relatively simple to modify, this is where many people start. Almost all performance

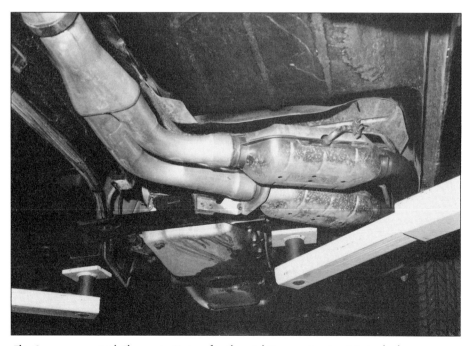

This Camaro uses a trick Flowmaster Y-pipe after the catalytic converters to minimize back pressure and make a smooth transition to the single 3-inch pipe.

A good high-flow muffler is also an essential ingredient in a high performance exhaust system. This Flowmaster muffler provides less back pressure than stock mufflers and has one of the "baddest" sounds of any muffler on the market.

exhaust systems available for the Camaro feature larger diameter tubing from the rear of the catalytic converter(s) and incorporate a high flow, performance muffler. Depending on the system, increases of up to 10 horsepower can be achieved with one of these aftermarket systems.

High flow mufflers use various techniques to lower exhaust back pressure and yet quiet the noise level significantly. There are a number of manufacturers that offer these exhaust systems, and you should examine them carefully (before buying) to determine if the material and tubing diameter are acceptable. Some are manufactured from the same type of steel tubing used on the factory exhaust while others are made from stainless steel. The stainless mufflers and exhausts are considerably more expensive, but they will not rust or crack after a long period of use.

Camaros from '82 and '83 will benefit from upgrading to a 3" diameter system. Later models (from '84-'92) already have a 3" diameter system, so all that remains is to install a low-restriction muffler in place of the stock one.

Chevrolet offers a free flow exhaust pipe/muffler package that features a 3" diameter stainless steel pipe and a low-restriction stainless muffler. Chevrolet claims this system delivers an extra 10 horsepower and is not significantly louder than stock. This system is sold as part numbers:

12341411: '84-85 305 TPI or HO

12341412: '83-87 305 (carb) and '88-90 305 EFI

12341413: '89-92 TPI engines with dual converters

12341414: '86-87 305 HO and '86-90 TPI with single converter

Aftermarket Mufflers—Dynomax, Thrush, Borla and Flowmaster also offer excellent low-restriction mufflers for third-generation Camaros. Systems from these companies typically feature 3" diameter, bolt-on exhaust pipes and a muffler to match. The Flowmaster muffler uses a series of chambers that the exhaust flows around to lower the noise and reduce back pressure. Borla manufactures a stainless steel system similar to the Chevrolet version and is

This 1989 Camaro is equipped with a 3-inch exhaust and a Flowmaster muffler. Installation of this type of system is simple and doesn't require any special tools or welding.

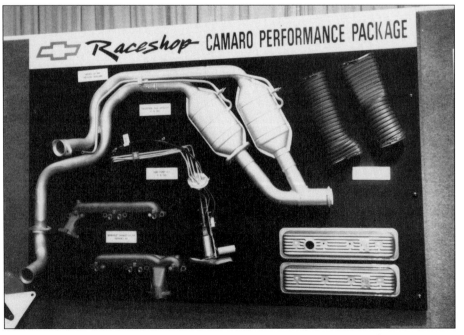

Chevrolet offers a complete, emissions-legal performance exhaust system for third-generation Camaros. It is usually on display at trade show and special exhibits.

slightly quieter than a Flowmaster muffler. The Borla muffler is almost straight through and uses perforated tubing surrounded by stainless fibers to control sound and reduce back pressure.

The Dynomax mufflers have proven to be very popular with Camaro owners. The Dynomax Super Turbo muffler and "Cat Back" system provides a throaty sound while offering little restriction to exiting exhaust gases. This muffler uses a system that reflects the sound waves back toward the inlet and absorbs them.

Although the "dual exhaust" look, with a tailpipe peeking out from beneath both sides of the rear bumper is most popular, it is not the most efficient when a single muffler is used. If a muffler has dual outlets, with one positioned on the left and one on the right, the diameter of the inlet pipe must be reduced. As an example, the Super Turbo muffler supplied in the Dynomax Cat Back systems include a 3" diameter inlet and dual 2-1/2" outlets when both outlets are positioned on the same side. However, when the muffler includes both a left side and right side outlet, the inlet is only 2-1/2" in diameter. Flowmaster also offers a selection of mufflers with dual outlets on one side or a single outlet on both sides. As might be expected, inlet size is smaller when outlets are positioned on each side of the muffler.

Thrush has taken a slightly different approach to low-restriction mufflers for third-generation Camaros. In addition to the standard "dual exhaust" configuration, Thrush offers its CVX muffler with a 3" inlet and single 3" outlet. While this arrangement doesn't provide dual pipes at the back bumper, it does deliver improved efficiency.

The second part of Chevrolet's Camaro high performance exhaust system is this muffler package. Note the Flowmaster designed Y-pipe.

EXHAUST MANIFOLDS/HEADERS

Tubular exhaust manifolds, the legitimate name for headers, are probably one of the most commonly sought-after pieces of performance equipment. Headers increase horsepower by providing an engine's exhaust gases with a smoother, less-restrictive path than stock cast-iron exhaust manifolds.

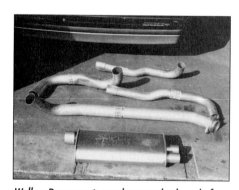

Walker Dynomax is another popular brand of high performance exhaust systems. This system (part no. 17499) features 3-inch exhaust pipes and a muffler with dual outlets on the same side. Photo by Dave Emanuel.

Dual exhaust, with a pipe peeking out from beneath the bumper on both sides may look good, but it's not the most efficient configuration, with a transverse-mounted single muffler.

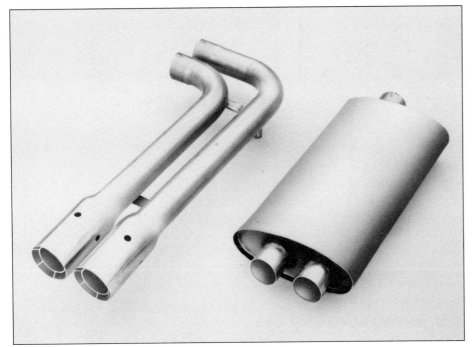

Borla's stainless steel muffler and exhaust pipes sound good, look good and will last almost indefinitely. The muffler is available separately or as part of a complete system.

If you can't afford a set of headers, these factory performance cast-iron manifolds, part #14094063 LH and #14094064 RH feature larger 2 1/4-inch outlets.

Choosing Headers

If you're on a limited budget, new aftermarket headers may be out of the question. In that case, the best option is to install the exhaust manifolds used on TPI engines (if your Camaro isn't already so equipped). These are available from Chevrolet as part number 14094063 (left) and 14094064 (right). These exhaust manifolds feature a 2-1/4" diameter outlet and can be used with 305 H.O. or 305 engines with throttle body injection.

For engines that have been modified to increase horsepower—such as those equipped with a performance camshaft, performance PROM chip, high flow muffler and aftermarket Tuned Port runners—a set of headers can yield as much as 30 extra horsepower. The next question then, is which type of header is best?

Tri-Y—As might be expected, that's a matter of opinion. One design that is particularly well-suited for use in a street-driven vehicle is the Tri-Y. This configuration starts off with four tubes that merge into two tubes that merge into one tube. As the name implies, there are three Y-junctions in this design. This type of header tends to deliver more bottom-end torque and horsepower up to 5000 rpm. That's ideal for late-model Camaro engines which make maximum horsepower below 5000 rpm.

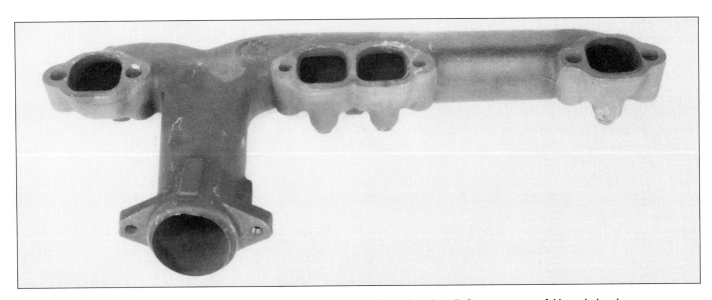

The performance factory manifolds installed on most '89-and-later TPI V8 models flow relatively well (for cast-iron manifolds) with their larger port openings.

Four-Tube—Other headers follow a four-into-one design where four exhaust tubes feed into one collector. This type of header is primarily intended for increases in performance at higher rpm levels. But with the limited space in a third-generation Camaro's engine compartment, and the need to bring both headers to a common collector before the catalytic converter, four-tube headers must be relatively short. Consequently, the pipes don't offer the higher rpm "tuning" advantage of a standard four-tube header.

Tubing Size—Headers also vary in tubing size as well as their design. Headers can range from 1-1/4" in diameter to 2" and larger. Cars with more horsepower will require larger diameter headers. For most performance applications however, even up to 400 horsepower, a standard set of 1-5/8" diameter headers are the best choice. This size of header will fit very easily in the cramped engine compartment of most '82-92 Camaros. Larger diameter headers are intended for higher output racing engines, but more importantly, they are more difficult to install and have a tendency to interfere with other components in the engine compartment that are normally removed in race cars.

Legality—The headers you choose must have an EO number, or must be EPA accepted for use on pollution-controlled vehicles like '82-and-later Camaros. As mentioned previously, use of headers that are not emissions-legal makes you subject to a $2500 fine, since they are

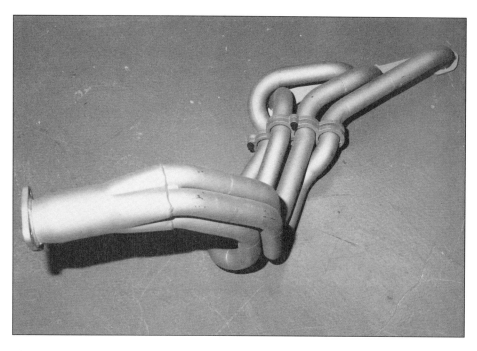

This custom race header system is from Arizona Speed and Marine and is made for large displacement small blocks and high rpm use. Space limitations require the two-piece construction which makes installation possible.

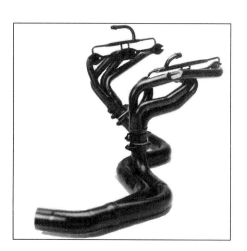

Emissions-legality also plays an important role in the header selection process. Systems like this one are available from Hedman and Edelbrock and are emissions-legal in all 50 states.

Anyone who has ever installed a set of headers knows that standard bolts are self-loosening. Stage 8 Fasteners offers stainless steel bolts and retainers which eliminate the need for periodic retightening.

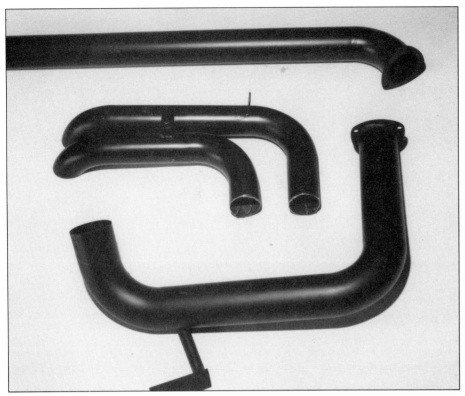

Part of a good exhaust system is the use of 2 1/4-in. or larger tubing. This 3-inch tubing system is available from Flowmaster and features mandrel bent tubes from the rear of the catalytic converter to the muffler.

clearly in violation of the Federal Clean Air Act. That's really just another way of saying that headers installed on third-generation Camaros must have provisions for installing an air pump and a fitting to accept an oxygen sensor.

Materials—Like the differences in design and tubing size, headers also differ in the materials they are manufactured from. Most late-model Camaro headers are made from either heavy gauge mild steel or stainless steel tubing. Mild steel headers are cheaper than their stainless counterparts, but also subject to corrosion. Most header companies offer special aluminized coatings which virtually eliminate problems with corrosion and rust.

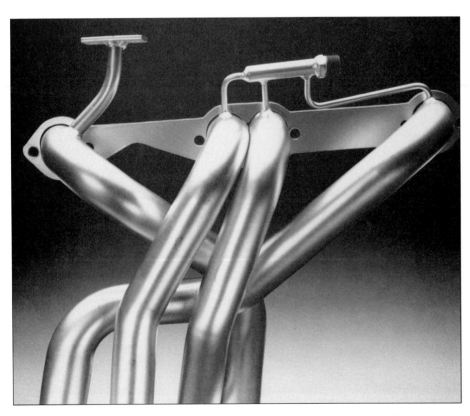

Headers come in various designs and sizes. This 1 5/8-inch diameter fou-tube header from Accel has a 3-inch collector and is intended for high output engines.

It may come as a surprise that Tri-Y headers are more expensive to produce than four-tube types. The Tri-Y design was widely used in the early Sixties but interest faded as cheaper four-tube headers became a comeback of sorts because they can be more easily fit into crowded engine compartments.

INSTALLING A PERFORMANCE EXHAUST SYSTEM

Test results from various manufacturers have shown that installation of a set of headers and a free-flowing exhaust can deliver as much as 25-30 extra horsepower. Prior to attempting the installation of headers, read the manufacturer's instructions completely and make sure you have all the necessary tools and equipment.

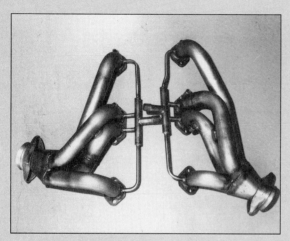

1. This stainless steel header system from Summit/SLP is designed specifically for late-model Camaros and is emissions-legal. This particular header features 1-3/4 inch diameter primary tubes.

2. The first step in installing a set of headers in the cramped Camaro engine compartment is to remove anything in the way. On this '91, we had to remove the dump valve and air injection assembly to gain access to the exhaust manifold bolts.

3. The air injection tubes should be carefully removed as not to bend or strip the tube fittings. A good flare nut wrench will to the job well and some WD-40 may be necessary to loosen the fittings. Once completed the exhaust manifolds can then be easily unbolted.

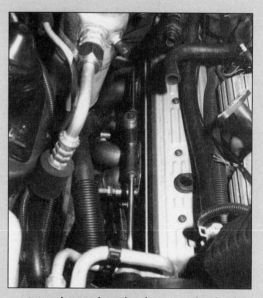

4. It took several tries but the Summit/SLP headers did slip into the correct position and were bolted in place. The driver's side is much easier since there is more room in that area.

5. The next step is to remove the factory Y-pipe. In single converter models, these pipes may be disconnected from in front of the converter. On dual-converter models, the pipes and converters must be removed.

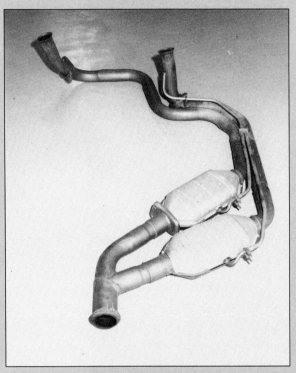

6. Dual cat systems are unbolted from the exhaust manifolds and a clamp at the rear of the converters. In this application, the tubes entering the converters will need to be cut.

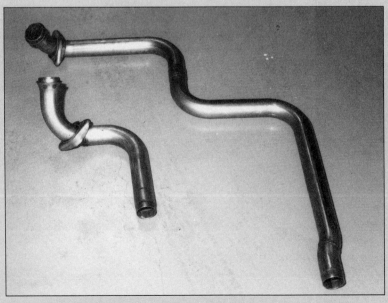

7. The supplied stainless tubes in this kit need to be welded in place of the factory tubes in front of the converters. Then, the entire assembly can be bolted back to the chassis.

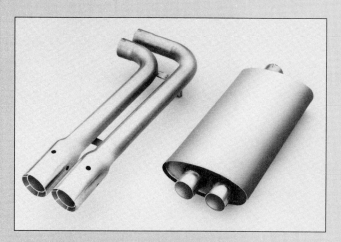

8. No free-flow exhaust system is complete without a good muffler system. This stainless muffler from Borla provides excellent flow with a healthy exhaust tone.

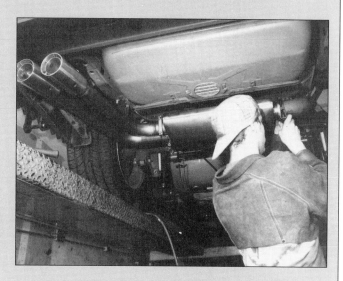

9. The factory 2-1/4 inch tubing from the catalytic converters' back can be used for this Borla muffler and it simply clamps into place and hooks into the stock hangars.

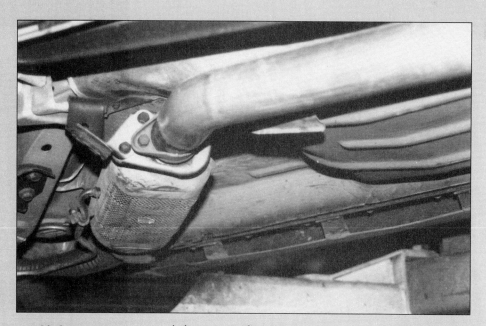

10. Some systems use a 3-inch diameter pipe from the rear of the converters. Obviously, larger diameter tubing is preferred for maximum flow capacity.

The TPI manifold is actually a four-part affair consisting of a throttle body, plenum, runner tubes and manifold base. This fully modified system includes a 52mm throttle body, ported plenum, large tube runners and TPIS "Big Mouth" manifold.

INDUCTION SYSTEMS 5

When it comes to setting up a performance induction system for a Camaro, there are a couple of factors that should be addressed. One is the cost involved and the second is how much extra air your engine requires. One of the most common mistakes on both fuel injected and carbureted engines is adding larger manifolds, plenums and carburetors that can actually hinder performance. As the saying goes, "bigger isn't always better."

The best way to increase your Camaro's performance is to modify the induction in stages, according to the needs of your engine. Considering the fact that there are a lot of performance aftermarket items such as throttle bodies, TPI runners, manifolds and plenums, it's very feasible to gradually increase performance. Most of these performance items work ONLY if the engine can take advantage of the increased airflow potential they provide. If your engine has modifications like a performance camshaft, ported heads and exhaust headers, it is obviously going to have a higher airflow requirement than a stock engine. However, even stock engines can benefit from modifications that improve the efficiency of the induction system.

The late-model TPI engine is one of the most sought-after engines for performance purposes. However, engine swaps can be tricky because 1989-and-earlier models incorporate a Mass Air Flow sensor, 1990-92 models do not. Each type of induction system requires a specific computer and wiring harness.

The best way to put an induction system in proper perspective is to picture a 4" diameter fire hose with short lengths of garden hose spliced in at several locations. Each piece of garden hose is a restriction, and as each one is exchanged for a piece of hose of larger diameter, overall flow capacity increases. But as long as a single piece of garden hose remains, a flow reduction will result. Full flow potential will not be realized until the entire length of hose maintains a diameter of 4".

If it isn't possible to maintain a consistent diameter from end to end, it just may be that hose diameter will have to be reduced in a number of locations. That being the case, the largest obtainable diameter will determine maximum flow capacity. Irrespective of changes made elsewhere, if it's necessary to include a length of 3" diameter hose, maximum flow capacity will be determined by that section of 3" diameter hose.

Keeping the hose analogy in mind it's apparent that performance won't increase much if a high performance TPI manifold base is installed and the runners and plenum are left stock. Similarly, porting the plenum without modifying the runners and manifold base will mean less than top performance.

AIR CLEANERS

The area where the air first enters into your engine is through the air cleaner, so that's the logical place to start. There are a variety of aftermarket air filters on the market which

An aftermarket 160-degree thermostat can help reduce induction temperatures and increase power slightly. However, if coolant temperature is too low, the computer won't go into closed loop mode and fuel economy and driveability will suffer.

Whether the intake system is stock or modified, a low-restriction air cleaner should be part of the program. Oiled cloth-type elements offer an additional advantage—they can be cleaned and reused.

improve airflow and have high filtering characteristics. Hypertech, K&N, TPI Specialties and Accel market low-restriction air filter elements for late model Camaros. The K&N filter is a good example of a gauze-type filter, that traps dirt in various areas of the filter element.

This type of filtering allows air to maneuver through dirt-blocked areas, so gauze filters provide relatively little restriction, even when they're dirty. By comparison, paper element filters keep dirt trapped on a level surface, where airflow is blocked, unable to circulate around the debris.

Accel is another manufacturer that offers low restriction, high efficiency air filters. Accel filters use the same method of dense filtering as the K&N filters. But instead of an oiled gauze material, Accel uses foam layers that can filter out small particles. The foam traps dirt in its denser structure and the air is free to pass around it, allowing for superior flow over standard paper element filters. Whichever type of filter you choose, this should be the first step to increasing your Camaro's performance.

Air Inlet

One simple, inexpensive modification may be performed on the air inlet into the engine. On carbureted

Underdrive pulleys don't have anything to do with the induction system, but they can result in a measurable horsepower increase because they reduce parasitic power loss.

This H.O. engine is fully emissions-legal and with the help of this dual-snorkel air cleaner, it can launch this Camaro to low 13-second quarter-mile times.

and throttle body injected engines, installation of a performance air filter will help, but the filter housing itself is also a restriction. The hot tip is to install a Z28 dual-snorkel air cleaner. These may be obtained at a salvage yard or purchased new from a Chevrolet dealer. The dual-snorkel air cleaner, which is part of the emissions-legal ZZ3 engine package offered by GM for 1982-1987 carbureted Camaros, is listed as part number 25042821. (The ZZ3 engine, part number 10185072, is the emissions-legal version of the 350 H.O. replacement engine. The engine is available separately, and it is the foundation of Chevrolet's complete emissions-legal package.) Also required are left and right snorkels (part numbers 14070917 and 14070918) and left and right side ducting (part numbers 14073299 and 14083990).

TPI Inlets—On TPI models, the air inlet is located under the front air dam of the car. The air then passes through a plastic collector and diverts into two chambers that house the square-shaped air filters. If you remove the filters, you'll see two baffles that can be removed to increase airflow to the air filters. For further improvements, you can drill up to six 1/2" diameter holes through each air cleaner box to allow even more air to enter. If you do this modification however, make sure that you use a washable, performance air filter like the Accel or K&N filters since this modification tends to collect more dirt. It also increases the possibility that water will splash up to the filters.

CARBURETED SYSTEMS

Although carbureted intake systems are not as exotic as the newer Tuned Port Injection set-ups, a late-model Camaro with a Rochester four-barrel carburetor has excellent horsepower potential—with the right combination of parts.

Carbureted 305 and 350 engines are somewhat easier to modify than their TPI counterparts. In fact, a carbureted engine has the potential to produce more horsepower than its TPI-equipped sibling, but it won't fare as well in the production of torque. The long runners of the TPI system restrict airflow at engine speeds above 4750-5000 rpm, and therefore bring an abrupt halt to the production of top end horsepower. However, below that speed range the runners are a definite asset and place a small mountain right in the middle of the torque curve.

Rochester QuadraJets

The stock Rochester four-barrel carburetor has been in use on Chevrolet vehicles for quite some time. It offers very reliable performance and can be fine-tuned relatively easily with the proper metering rods. Before attempting to make any modifications however, give the carburetor a good going over to make sure it is in proper working order. In addition to the normal mechanical parts, the electronic components—the Throttle Position Sensor and Mixture Control Solenoids—must also be in proper working order. If necessary, rebuild the carburetor and replace any non-functioning parts. HPBooks has an excellent guide, *Rochester Carburetors*, to help you with any Q-Jet project.

LG4 Quad—For performance applications, the Rochester carburetor used on LG4 engines, part number 17111478 is a good choice. Because Camaro carbureted engines are controlled by the computer system, it may not be possible to use an aftermarket carburetor. Emissions regulations require that a carburetor must be an original equipment replacement, EPA accepted or approved by the California Air Resources Board (CARB).

Fortunately, the LG4 QuadraJet is more than adequate for just about any street performance engine with a computerized engine management system. Typically, the only carb modification required is a recalibration of the secondary fuel metering system. For most applications, substituting a set of AH metering rods and a G hanger for the stock components will suffice. (These are the same rods and hanger used in carburetors that are installed on the ZZ3 engine package.) These parts should be available at any carburetor-oriented AC-Delco dealer.

Holley—Although Holley offers more carburetors than a politician has campaign promises (hard to believe), there is currently only one that is suitable for late-model Camaros. Holley's Spread-Bore replacement for use on 1982-83 305 engines is listed as part number 0-80073. It features vacuum secondary operation and retains full computer controls.

Carbureted Intake Manifolds

For further improvements, a standard intake manifold can be swapped for the later model H.O. aluminum manifold. Its relatively low profile makes it a direct bolt-on replacement for the stock, cast-iron unit. The H.O. manifold, part number 10185063, is used on the Chevrolet 350 H.O. replacement engine and provides good bottom-end torque and mid-range power.

Similar manifolds sold through Edelbrock, Weiand and Holley are also available, but many are not emissions-legal because they don't include provisions for some pieces

This H.O. engine is topped off with a low-profile aluminum intake manifold and computer-controlled Rochester carburetor. These components not only produce excellent power, they're emissions-legal.

This is an example of an excellent high performance intake manifold that is not emissions-legal. Even though it has a "Bowtie" emblem and an official Chevrolet part number, this manifold is designed for use on engines that were originally equipped with a square-flange Holley four-barrel.

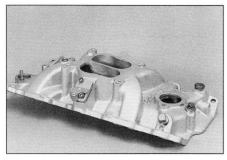

Edelbrock offers a variety of intake manifolds that are considered to be stock replacements. As such, they are fully emissions-legal, but offer better-than-stock performance. Performer part number 3701 may be used on Camaro engines originally equipped with a Rochester four-barrel.

of emissions-control equipment. Others are too tall and will place the air cleaner in a position that will "recontour" the hood when it's closed.

Both Edelbrock and Weiand offer four-barrel intake manifolds that meet emissions requirements and therefore can be legally installed on third-generation Camaro engines—provided the engine was originally equipped with a four-barrel carburetor. When shopping for an intake manifold, keep in mind that 1986-and-earlier cylinder heads require a different manifold than 1987-and-later cast-iron heads. The later heads have their two center manifold bolt holes drilled at a different angle than earlier heads. It is possible to modify a manifold to fit, but it's a lot easier to start with the right parts.

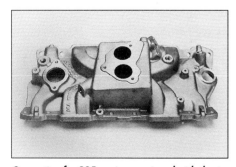

One option for 305 engines equipped with throttle body injection is the installation of an Edelbrock TBI intake manifold. This stock replacement-style manifold is completely compatible with the stock throttle body.

THROTTLE BODY INJECTION

The late-model GM throttle body injection system is more efficient than a carburetor and cheaper to produce than Tuned Port Injection. Consequently, it has become the standard fuel delivery system for base Camaro engines. The throttle body injection unit looks somewhat like an electronic carburetor, but rather than requiring engine vacuum to draw fuel into the air stream, it injects fuel under pressure.

Throttle body performance has definitely increased in the past few years and there are many late-model Camaros with V6 and V8s equipped with this system. The aftermarket

Turbo City TBI improvement—the Air Flow Enhancer—replaces the stock air cleaner base and increases airflow by as much as 10%.

industry is starting to respond with items, recognizing that the TBI system can be made to produce as much power as the more expensive TPI system with some modifications. Edelbrock (2700 California Street, Torrance, CA 90503 310-781-2222) is one manufacturer that makes a performance throttle body intake manifold and offers a replacement camshaft to match. Crane Cams also offers a camshaft unit that reportedly will boost horsepower by as much as 15 percent.

Turbo City (1137 W. Katella Ave. Orange, CA 92667 714/639-4933) has developed a line of products to enhance the stock TBI system as well. Their items range from complete packages, to individual bolt-on items such as an *Air Flow Enhancer*, which replaces the more restrictive stock air cleaner base in the throttle body, increasing airflow by 10%. Another item is their TBI spacer, which functions much the same way a carb spacer does by elevating the TBI unit. This increases airflow potential, and improves fuel atomization. Dyno testing has shown that the spacer alone will produce as much as a 12 hp increase. Turbo City has also developed replacement TBI injectors that are flow-matched for consistent flow.

The best performance from throttle body-equipped V8 Camaro engines has been obtained through the use of the Edelbrock manifold with ported, aluminum Corvette heads, a performance camshaft and headers. The important thing to remember here is that all of these items create more power when used together. Simply changing the manifold or camshaft alone will not deliver the type of horsepower increase that can be obtained from a combination of coordinated components. Make sure you consider the entire package, not just the individual parts.

Turbo City offers several TBI performance products, including these injectors which have been flow-matched to within 1% of each other. Stock injectors have been known to vary as much as 15%.

57

Some Camaro owners are more concerned with appearance than performance. The induction components on this engine have been given the full chrome treatment.

TUNED PORT INJECTION

Tuned Port Injection has always attracted performance enthusiasts so naturally, it has received the most attention from manufacturers of high performance parts. Tuned Port Injection is actually a proprietary name for a unique form of multi-port fuel injection. This particular system employs long, curved manifold runners that are tuned to provide maximum power in the middle rpm range—from approximately 2000 to 4000 rpm.

Chevrolet did an excellent job engineering the TPI system, so large horsepower gains won't be realized from the installation of a single component. Rather, a systems approach is required; the key to significant power gains is to install a number of components, each of which contributes to the overall gain.

Air Foils

Although there is some debate as to the effectiveness of installing an air foil in the flat area between the throttle bores, you need only listen to a TPI throttle assembly on a flow bench to know that the foil has a definite effect. With no foil in place, air rushing through the throttle openings creates a harsh loud sound, a result, no doubt, of turbulence. With an air foil in place, the sound is noticeably quieter and smoother—it sounds like the air wants to flow through the openings rather than fighting to stay out.

Flow bench tests bear this out as the typical TPI throttle body assembly flows 40 cfm more air when an air foil is installed. However, when an air filter, Mass Air Flow (MAF) sensor and ducting are connected to the throttle body and the entire assembly flow tested, installation of an air foil increases flow by only 16-17 cfm. The difference is attributable to the restrictions imposed by

A modified intake system can pay off with significantly improved drag strip performance. With Edelbrock's TPI runners and manifold, this All Chevy magazine project Camaro ran almost a full second quicker in the quarter mile.

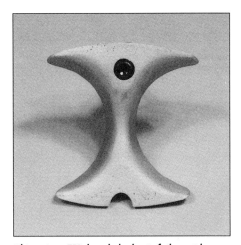

This unique TPI throttle body air foil provides a small, but noticeable increase in power. Air foils also seem to increase fuel economy and throttle response.

engine, the air foil will have increasingly more effect.

TPI Runners

High performance TPI runners seem to be popping up everywhere. And as might be expected, some models perform better than others. In all cases, high performance runners are larger in diameter than their stock equivalents and consequently have a higher airflow potential.

That increased airflow potential translates into increased torque and horsepower and also raises the torque peaks 250 to 500 rpm. Along with that comes a slight decrease in power below 2000 rpm. Installation of aftermarket runners is most effective if the engine has already been equipped with headers and a low-restriction exhaust system.

Port Matching—Be forewarned however, that TPI runners can actually hurt performance if they're not port-matched to the plenum and manifold. In the rush to bring TPI products to market, or to gain a

Several options exist for oversized runners. Some are constructed with tubing, others are cast as a single piece. There doesn't appear to be any clear-cut advantage to either type except that tubing runners are typically larger.

the MAF sensor, filter and ducting; by itself, a TPI throttle body assembly will flow approximately 710 cfm with an air foil installed. With the filter and MAF in the system, maximum airflow drops to 585-590 cfm.

It should also be noted that smaller, mildly tuned engines won't benefit as much as their larger, more aggressive relatives. So a stock 305 won't respond as enthusiastically as a modified 350. Irrespective of displacement or level of modification, installation of an air foil almost always results in improved throttle response.

The air foils marketed by various manufacturers are very easy to install, relatively inexpensive and legal in all 50 states (see sidebar nearby on how to install TPI Specialties air foil). As such, installation of an air foil is a good starting point for TPI system modifications. However, don't be disappointed if after installing an air foil, you don't feel a difference in acceleration. It probably won't increase horsepower enough to be noticeable. But you should notice a crisper and quicker response from the engine when you step on the throttle. And as you add more performance items to your TPI

price advantage, some manufacturers neglect proper engineering. Runners that aren't correctly designed and manufactured will cause poor idling and a lack of performance at higher rpm. Both the siamesed and large individual tube runners from Arizona Speed and Marine, TPI Specialties and Lingenfelter are excellent products. However, if aftermarket runners are bolted to a modified plenum or manifold, additional port matching may be required as most models are matched to stock dimensions.

Siamesed runners are another option for TPI-equipped engines. As the name implies, the two adjacent runners are opened up to a common larger one. This arrangement provides maximum airflow, but reduces low-speed torque.

Edelbrock offers a high-flow base and large tube runners for 305 and 350 TPI engines. In addition to increasing power, these emissions-legal components offer a unique appearance.

TPI Manifolds

Larger runners naturally go hand-in-hand with a higher capacity intake manifold; it's best to purchase and install both at the same time. There are several TPI manifolds on the market. As of this writing, only the TPI Specialties "Big Mouth" manifold, Edelbrock TPI manifold and the Accel/Lingenfelter manifolds are emissions-legal.

If the price of a new manifold is beyond your budget, it's possible to have the stock manifold ported. This can be done through conventional means or with a process known as

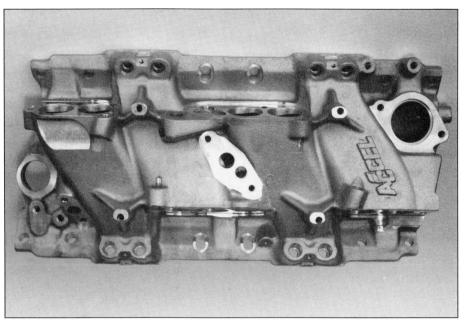

Accel's TPI manifold base has improved runner flow and larger port areas. It is recommended for high performance engines with cam and head modifications or larger displacement stroker engines such as 383, 406 or 427 cubic-inch small blocks.

Extrude Honing which uses an abrasive putty to remove material. Runner diameter can also be increased through the Extrude Honing process, but the amount of material that can be removed is limited. Aftermarket runners have a higher airflow and horsepower potential.

Plenums

The TPI plenum is often overlooked, but it is one of the most important links in the horsepower chain. The plenum is basically a storage area for air, but it also affects airflow potential. If the stock TPI configuration is retained, the only viable approach is to port the stock plenum. Aside from removing flow restrictions, the purpose of plenum porting is to increase volume so more air can be stored. Ported plenums are available from a variety of sources, however, with a high-speed die grinder, some cartridge rolls and a steady hand, plenum porting can be a do-it-yourself proposition. The primary goal is to remove any sharp edges and radius and blend the openings to which the runners connect. Pacific Performance in Huntington Beach, CA, has found that the Accel/Lingenfelter Super Ram Plenum works tremendously well on modified 305 and 350 engines. Ported Super Rams also promote good horsepower and increased torque in 383 or larger "stroker" engines.

Supposedly, there's no truth to the rumor that the "Big Mouth" manifold from TPI Specialties was named after company president Myron Cottrell. However, it is true that this manifold has been granted a CARB exemption, making it fully emissions-legal. In combination with large tube runners and a ported plenum, this manifold delivers an impressive power increase.

Accel's "box" manifold was designed by John Lingenfelter. It mates to special runners and works well with modified engines.

A recent *All Chevy* Magazine engine project used a Lingenfelter Super Ram on their 434 cid small-block stroker. The result was 550 horsepower and 600 lbs-ft. of torque!

TPI Throttle Bodies

The TPI throttle body should not be confused with throttle body injection. All throttle bodies contain the throttle plates that control the amount of air entering the engine. Throttle body injection systems have the injectors mounted above the throttle plates, so the throttle body controls the flow of fuel and air.

In a multi-port injection system, such as Tuned Port Injection, the throttle body assembly controls only airflow, because individual injectors are mounted in each manifold runner. The TPI throttle body also houses the throttle position sensor and the idle air control solenoid.

It seems obvious that a larger throttle body increases airflow capacity and that should increase horsepower. But think back to the fire hose analogy. If there's a significant restriction somewhere else in the induction system, a throttle body with larger openings will have little, if any positive effect. It makes sense to install a larger throttle body only if an engine has the need for it. In general, that need will arise only if a 350 cid or larger engine has been treated to aftermarket runners and manifold, a ported plenum, headers and modified cylinder heads.

Several manufacturers such as TPI Specialties, Pacific Performance,

Larger air requirements call for larger throttle bodies such as this 1000 cfm unit from Accel. It will fit stock and aftermarket plenums. However, it's a bit of overkill for a stock engine.

Accel, Lingenfelter and Arizona Speed and Marine offer a selection of throttle bodies with throttle bores ranging from 52 mm to 58 mm. Depending on the size, flow capacity can range from 650 cfm to 1000 cfm.

The best way to select the proper size throttle body is to contact the manufacturer and discuss the types of engine modifications that you have or will make in the future so they can recommend the proper throttle body size.

Fuel Pressure Regulators

The availability of adjustable fuel pressure regulators has caused some confusion among late-model Camaro enthusiasts. As delivered, most TPI engines have fuel pressures ranging from 35 to 47 psi, depending upon model year. With the stock pressure regulator being a mass produced item, it's common for actual fuel pressure to vary significantly from the factory specification.

Installation of an adjustable pressure regulator not only provides the capability to establish specified fuel pressures, it also allows for experimentation at other pressure settings. But before attempting to install an adjustable pressure regulator, it's

In addition to the "Big Mouth," TPI Specialties also offers the Mini Ram, which is very similar to the LT1 intake manifold. Although not yet CARB exempt, the company is in the process of filing for approval.

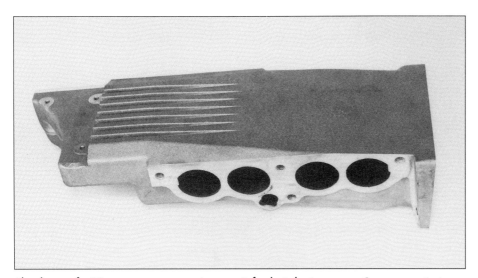

The plenum of a TPI system serves as an air reservoir for the induction system. Storage capacity is increased and flow restrictions minimized by porting and matching the outlets to runner diameter.

necessary to check pressure in stock condition. Also note that system pressure should be relieved before working on any part of the fuel delivery system. This is easily accomplished by depressing the center pin of the Schrader valve located on the passenger-side fuel rail.

At first glance, it may seem that raising fuel pressure would serve no purpose, because the ECM will simply sense (through the oxygen sensor) a richer mixture and alter injector pulse width to compensate. Although that is precisely the action taken, higher pressure is advantageous because fuel atomizes better. In fact, it's not uncommon for a knowledgeable engine builder to install downsized injector nozzles specifically so that higher fuel pressures can be run without causing an excessively rich condition at maximum fuel flow levels.

So long as fuel pressure is kept below 55 psi, the ECM should have no problem making the pulse width modifications necessary to keep air/fuel mixtures within the ratios programmed into the system. That's one of the most appealing aspects of computerized engine controls—to a large degree, the system is self-compensating. If an increase in fuel pressure richens the intake mixture, the ECM narrows injector pulse width to bring the air/fuel ratio back to the ideal 14.7:1 (stoichiometric); if pressure is lowered, the ECM moves pulse width in the opposite direction and air/fuel ratio again returns to 14.7:1 (under part throttle operation).

Problems arise only at or near wide open throttle. Under these conditions, a richer 11.5:1 air/fuel ratio is commanded, but the system does not read the oxygen sensor. Consequently, if injector capacity is mismatched to an engine's fuel requirements, mixture at wide open throttle can be excessively rich or lean. However, with a scan tool or Diacom, this is easy to detect by simply looking at oxygen sensor voltage.

Injector Nozzles

In theory, all injector nozzles of a rated capacity are created equal. In actuality, they often are not; flow capacity can vary 20% among the eight nozzles found in a TPI manifold. Since new injectors are relatively expensive, the most economical way to achieve a balanced fuel flow capacity is to have an existing set of nozzles flow checked and blueprinted. TPI Specialties, among others, offers this service.

Although small variations in injector flow capacity might not seem significant, they can be surprisingly so. In most TPI engines, there's only one oxygen sensor so the ECM looks at only one side of the engine to determine whether mixtures are rich or lean. If an exceptionally lean injector is located on a right bank

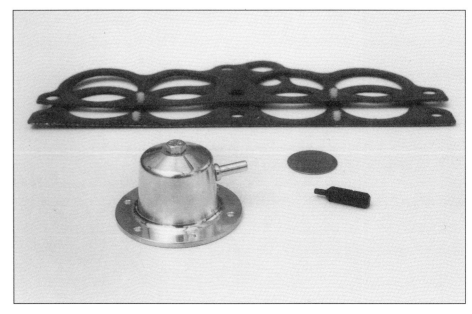

The Torx bolts that hold the pressure regulator in place are a special "security" type and require a specific wrench. It's included with adjustable regulator kits.

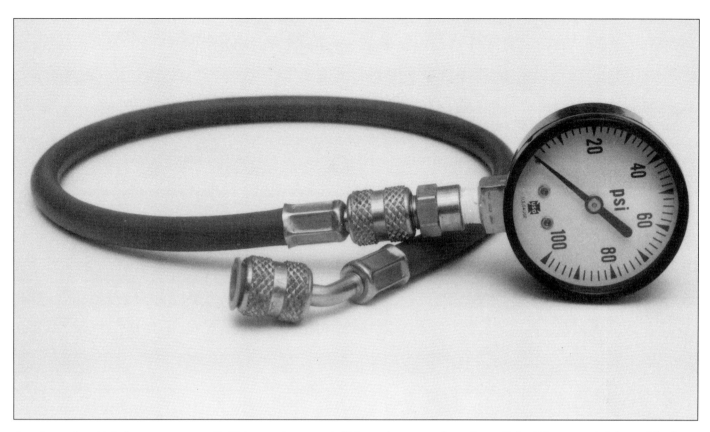

Experimentation with fuel pressure can pay off with increased power and efficiency. A gauge can be easily installed on TPI engines which have a Schrader valve on one of the fuel rails.

cylinder, and the oxygen sensor is on the left side, it will never see that condition. If a right side cylinder is lean enough, it may cause detonation whereupon the knock sensor will enter the equation, timing will be retarded, and maximum power will decrease.

In most instances, existing injector nozzles can be flow balanced and matched. However, some nozzles may be too far gone to be brought up to specifications. If replacement injectors are required, they can be obtained from GM dealers or from independent fuel injection specialists. MSD Ignitions offers a wide selection of GM Multec injectors, which are available through an extensive dealer network—usually at lower cost than through other sources. Stock replacement injectors for 305 TPI engines are rated at 19 lbs-hr fuel flow; replacement injectors for 350 engines are rated at 22 lbs-hr.

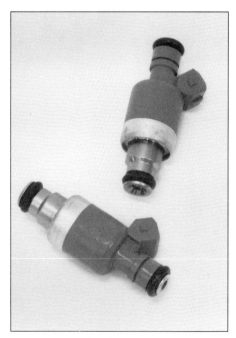

Large engines with high-output induction systems may require higher flowing fuel injectors such as these from Autotronic Controls. Injector flow rates range from a stock 305's 19 pounds per hour to 50 pounds per hour for high output replacements. However, bigger isn't always better.

Although the nozzles used in electronic fuel injection systems are precision components, flow rates can vary and the accumulation of deposits also alters flow rate. Companies like TPI Specialties have the equipment to flow check and blueprint nozzles for consistent fuel flow.

TPI POWER SECRETS

by Dave Emanuel

If the thought of disassembling a TPI system causes you to break out in a cold sweat—or worse yet, consider buying a Mustang— put your fears on ice. Once you get over the initial concern over of spinning your wrenches in uncharted territory, you'll find that the TPI hardware is no more difficult to deal with than a carburetor-based induction system. Reaching some of the bolts is indeed challenging, but once you've figured out the ratchets, extensions and sockets required, you're ready to rock and roll.

The following modifications can all be accomplished in a few hours, making a TPI hop-up an ideal project for a Saturday afternoon. The modifications pictured here involve installation of a TPI Specialties Fast Pak and a few ancillary modifications. Fast Pak kits are available from TPI Specialties, 4255 County Road 10 East, Chaska, MN 55318, 612/448-6021.

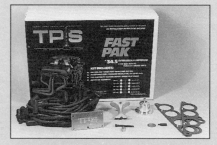

The TPI Specialties Fast Pak for TPI engines includes throttle body air foil, adjustable fuel pressure regulator, low-restriction air cleaner, high performance spark plug wires and a TPIS identification plate. It's one of the easiest ways available to hop up a TPI system.

1. Begin by removing the clamp that holds the ducting in place on the throttle body. The ducting and its attachment method differ according to model year.

2. It's not necessary to remove the throttle body from the plenum to install the air foil. Leave it in place so you don't have to wrestle with it while you're working. However, the four throttle body attaching screws will have to be pulled out later so the plenum can be broken loose, a requirement for installation of the adjustable fuel pressure regulator.

3. To install the air foil, remove the six Torx head screws that hold the cover plate in place.

4. The air foil nut (arrow at right) slips right into a small hole between the throttle bores. An Allen wrench is used to tighten the bolt that threads into the nut and holds the air foil in place.

5. Hold the nut in place and put the air foil in position. Then thread the retaining screw into place and tighten it. Replace the cover plate.

6. Four Torx-headed bolts hold each runner section to the plenum. After removing the throttle body, remove the runner attaching screws from each side. It may be necessary to loosen the bolts that hold the runner sections to the manifold and lightly rap the plenum with a rubber hammer to break it loose from the gaskets. But before you do that...

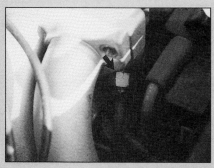

7. ....disconnect the fitting that attaches the power brake vacuum line to the plenum and...

8. ...remove the three bolts that hold the cable bracket to the plenum body. The cables can remain attached because only the plenum must be lifted out—the throttle body can remain in place.

9. With everything else disconnected, the plenum may be lifted, but can only be moved a few inches. The wire that plugs into this fitting on the bottom of the plenum (Manifold Air Temperature sensor) must be disconnected so the plenum can be swung to the side.

10. With the plenum flipped over as shown, you have all the access you need to the fuel pressure regulator. Don't forget to reconnect all electrical and vacuum lines during reassembly.

11. The stock pressure regulator is held in place by six security Torx screws (note small tang in the center of each screw head—that's the security feature). Special wrenches are available from a variety of sources and are included in TPI Specialties adjustable regulator and Fast Pak kits.

12. Before removing the stock pressure regulator, remove the gas cap (to eliminate pressure in the tank) and bleed off system pressure at valve/gauge connection. This valve functions like a tire valve. As you can see, the fuel pressure adjustment screw can be easily accessed after the system is reassembled. If you're smart, you'll check system fuel pressure before making any changes so that you can reestablish the baseline after the adjustable unit is installed.

13. Once the adjustable pressure regulator is bolted in place, the vacuum hose can be attached and the plenum reinstalled. Use new plenum/runner gaskets which are supplied with the adjustable regulator.

14. If your TPI system is equipped with a Mass Air Flow sensor, it has screens on each side. Flow capacity can be improved by removing these screens.

15. The easiest way to remove the screens is to roll a knife blade around the seam between the screen retainer and the MAF sensor body. Once you cut through all the way around, the retainer will pop off. Finish the TPI hop-up off by reassembling the plenum to the runners and throttle body and reconnecting the electrical and vacuum lines. Then replace the stock air cleaner with a low restriction model and swap the stock plug wires for some high performance helically wound type.

Although Corvette and aftermarket aluminum cylinder heads offer significant increases in airflow capacity, the stock cast-iron heads aren't all that bad. Road racing Camaros demonstrate surprising horsepower when equipped with stock cast-iron heads.

CYLINDER HEADS & VALVETRAIN MODIFICATIONS

When it comes to building a performance engine, the best approach is to build a solid, short block and save your money for the cylinder heads. A set of well-prepared heads is the key to impressive amounts of horsepower without sacrificing cost or driveability.

Many enthusiasts however, think the camshaft is the secret to making horsepower. This is true to a certain extent, but a proper set of heads is required before a camshaft can deliver its full potential.

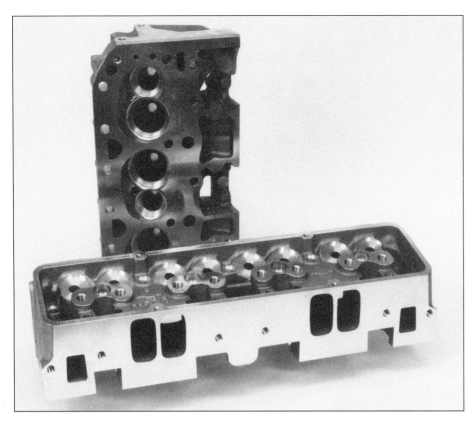

Chevrolet Bowtie heads are a bit too much for the engine in a street-driven Camaro. These are race heads that are designed for high rpm operation.

CYLINDER HEADS

Cylinder heads enable the optimum amount of air and fuel to reach the cylinders. The more air and fuel the engine can burn quickly and efficiently, the more power it is going to make. For example, larger intake and exhaust valves allow more air and fuel to enter and exit an engine than their stock counterparts. Heads with larger ports have more intake and exhaust port areas and provide even more horsepower gains, especially in conjunction with larger valves.

But there's an old saying that should be heeded—"Bigger isn't always better." This is especially true with 305 engines, because they don't demand as much air as a 350 and their small diameter cylinders (3-3/4" vs. 4.00") limit maximum practical valve size.

The use of ports and valves designed for larger displacement or extensively modified engines is ill-advised in a basically stock 305 or

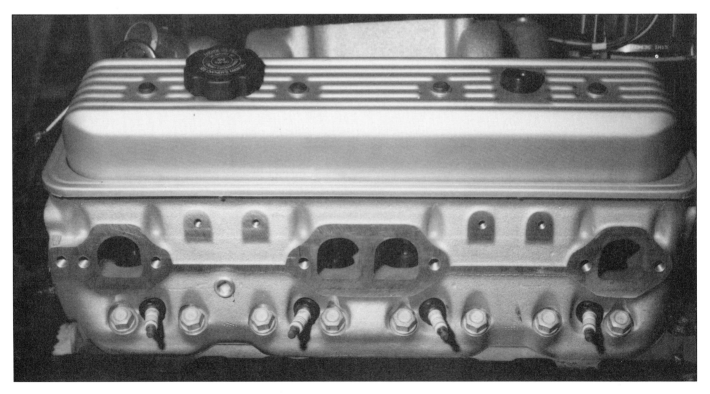

The aluminum heads found on the crate H.O. engines feature a D-port exhaust head design with angled spark plugs. These heads, which are identical to Corvette aluminum heads, offer improved airflow capacity and bolt to a stock Camaro engine with no modification. Of course, a little porting doesn't hurt.

The key to horsepower is airflow and the key to airflow is the cylinder head. When Air Flow Research obtained a CARB exemption for its aluminum street heads, it spelled good news for owners of late-model Camaros in search of more horsepower.

350. Large ports are designed for high rpm operation; at lower engine speeds, the relatively low airflow demands of a mild engine will cause the air and fuel to travel through the head very slowly. This lack of velocity can cause a stumble upon acceleration and lead to decreases in power throughout the rpm range. Again, the key to improved performance is careful selection of parts which can be assembled into an ideal combination.

So how does someone get started on achieving any significant gains in horsepower with the cylinder head? First, the volume of the ports can be increased for improved breathing capabilities. Normally this is done by porting and polishing the heads to increase their intake and exhaust port sizes and minimize restrictions to airflow.

Preparing a cast-iron cylinder head with these modifications, normally takes hours with a hand grinder, cutting and polishing away the walls of the head's intake and exhaust ports a little at a time. The amount of work involved to make these modifications, however, depends on the design of the head you start off with.

Head Swapping

Knowing that the stock Camaro heads ('82-'92) are not designed with optimum horsepower in mind should be a clue that considerable improvements can be made. A set of Corvette aluminum heads, or Air Flow Research emissions-legal aluminum heads is obviously the hot set-up.

Corvette Heads—Corvette heads can be obtained from various salvage yards, swap meets or new from the local Chevrolet dealer as part #10185086. These heads feature a 58cc combustion chamber and have a 163 cc intake runner volume. They also use standard 1.94" intake and 1.50" exhaust valves. In stock form, when installed on a 350 cid Corvette engine, these heads result in a 9.5:1-10:1 compression ratio. Finding a set of these heads at a salvage yard is not too difficult. Look for casting number 10088113 and check prices before plunking down any hard cash. Also, make sure the heads are returnable if any cracks or other damage is found.

AFR Heads—Air Flow Research's (Pacoima, CA) street-legal aluminum

Large port heads increase maximum airflow potential which leads to higher horsepower. However, combustion chamber volume must be correct. If the chambers are too large, compression ratio will nose-dive and power output will drop like a rock. These high performance heads from Air Flow Research are emissions-legal and available with just about any combustion chamber volume desired.

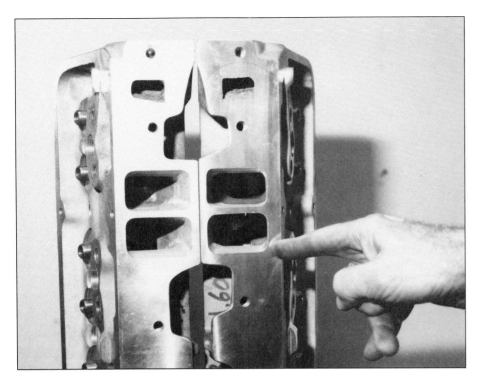

At first glance, many small-block heads look virtually identical. However, closer examination often reveals significant differences in port size. Note the difference in the amount of material above the ports in the two heads pictured. For best performance, it's best if port dimensions closely match those of the intake and exhaust manifolds.

cylinder heads are also a popular option. Although Air Flow Research (AFR) heads can be ported, they deliver exceptional performance right out of the box. If new cylinder heads are in the program, the AFR castings are the best way to go if you're attempting to make serious horsepower. Corvette aluminum heads do offer excellent performance potential, but they are, after all, designed for a stock production engine. Consequently, they require modifications to provide the same airflow capacity as a set of out-of-the-box Air Flow Research castings. AFR heads are supplied with 2.02" intake and 1.60" exhaust valves, as opposed to the Corvette head's 1.94"/1.50" combination, so they also offer higher overall flow capacity on two accounts—larger valves and larger ports. On the other hand, Corvette heads are a bit cheaper and are more than adequate for many applications.

Porting

Regardless of the castings selected, the approach to porting is the same. For a street engine, extensive porting is not required. All that's required is blending and enlarging the bowl or pocket area directly beneath the valve. Known as "pocket porting" this operation removes the sharp edges where the machined area transitions to the as-cast surface. Typically, the valve guide is also thinned and blended and the overall size of the pocket is increased. Beyond this, matching the ports to the intake or exhaust manifolds will complete the job.

Shopping Tips

When shopping for cylinder heads, keep in mind that Chevrolet made a change in valve cover configuration for the 1987 model year. Later model heads feature a raised valve cover rail and a string of valve cover bolt holes in the center of the casting. Older heads have the traditional bolt holes adjacent to the valve cover rail. Unfortunately, the changeover seems to have taken quite a while because both types of

Note the difference in port size and higher intake angles from the 198 cc head on the right to the 210 cc head in the center and the 220 cc head on the left. Photo courtesy of All Chevy magazine.

1987-and-later small block heads include a slightly different angle on the two center intake manifold bolt holes. When purchasing heads, make sure that the bolt hole angles match those in the intake manifold that will be used. Photo by Dave Emanuel.

heads turn up in "incorrect" applications. Problems will therefore be minimized if the currently installed heads are properly identified and the replacement heads have the same configuration. This will ensure that the intake manifold and valve covers fit correctly. Also, make sure that combustion chamber displacement of both the existing heads and their replacements are known. Otherwise, performance may be compromised by a compression ratio that's either too high or too low.

V6 HEADS

Chevrolet is the only source for performance heads for the 2.8 liter and 3.1 liter, 60-degree V6 engines. Chevrolet part number 14054884 is a cast-iron cylinder head that's used on High Output engines in both transverse-mounted (front-wheel-drive) and rear-wheel-drive applications. This casting is machined for 1.72" diameter intake and 1.42" exhaust valves (standard valve sizes are 1.60" intake and 1.30" exhaust). Obviously, a slight bump in compression ratio, installation of a high performance camshaft and some porting work are useful in taking full advantage of the head's airflow capacity.

The 1987-and-later "Generation II" engines are equipped with a canted-valve aluminum head (part number 10048649) that also has 1.72" intake and 1.42" exhaust valves. Although this head is appealing, it's not practical to install on 1986-and-earlier engines.

VALVES & VALVETRAIN

A good set of heads also need a proper set of valves to accommodate the larger increase of air and fuel entering and leaving the combustion chamber. The standard intake and exhaust valves in 350 cid Camaro engines measure 1.94" and 1.50" respectively for intake and exhaust. Moving to the 305 cid engines, intake valve size shrinks to 1.840". The 60-degree V6 engines are fitted with either a 1.60"/1.30" or 1.72"/1.42" intake/exhaust valve combination depending on model year. For most high performance street conditions, the most practical

71

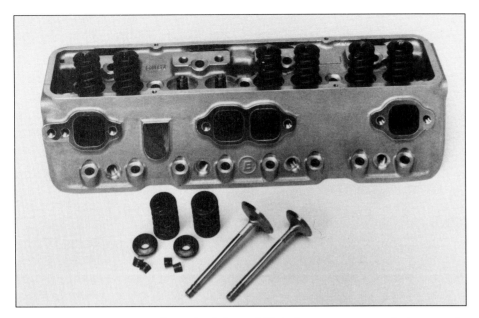

Edelbrock has released a series of street heads for small-block Chevy engines to cover both street and race applications. Some versions are emissions-legal.

approach is to install slightly oversized replacement valves. The investment required to accomplish a significant increase in valve diameter is justified only for race applications or in cases of terminal V6 infatuation.

As noted, installation of late-model aluminum Corvette heads on V8 engines will yield substantial gains in horsepower. However, these heads are equipped with 1.94" intake and 1.50" exhaust valves. Slightly larger intake valves (2.00") can be installed without any modifications, but new valve seats are required to move up to a 2.02"/1.60" combination. It should also be noted that it isn't cost-effective to install larger-than-stock valves in cylinder heads destined for 305 engines; the small bore shrouds the valves, thereby restricting airflow.

Options

Whenever new valves are required, it pays to install swirl polished high performance types such as those offered by Manley, TRW, Speed-Pro, Ferrea or SI. By virtue of their shape and finish, high performance valves promote higher airflow rates than their stock counterparts. Even greater flow capacity can be achieved by installing valves with undercut or "necked down" stems. Standard small-block Chevy valves have 11/32" diameter stems; when the lower portion of the valve stem (which passes through the port) is machined to 5/16" diameter, airflow increases noticeably, especially at lower valve lifts. Valves with undercut stems are usually machined from stainless steel and are available in 1.94" and 2.02" intake sizes and 1.50" and 1.60" exhaust sizes.

1987-and-later Camaros use a billet-type roller camshaft (top) with essentially the same profile as the hydraulic flat-tappet cam that preceded it. Part number #14093643 offers .404" intake and .415" exhaust lift; duration at .050" lift is 202 and 207 degrees. The 1988-89 cam has slightly more duration (207/213-degrees at .050" lift) and lift of .415" and .430".

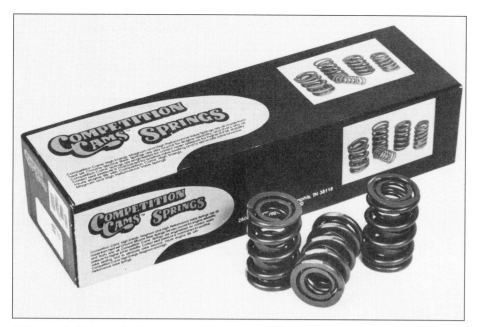

Roller camshafts require higher pressure springs to control lifter motion. Camshaft manufacturers like Competition Cams offer a wide selection of high performance springs.

race engines with cams having more than .500" lift.

Valve Springs

Along with high lift camshafts comes the requirement for higher pressure valve springs. Depending on the camshaft, the valve spring seats may have to be machined to accept larger diameter valve springs. Valve spring seat height may also need to be adjusted so that proper installed height can be achieved. Most mild high performance camshafts for late-model engines can be matched to stock diameter (1.25") valve springs, in which case seat machining will not be required.

Valve Job

Regardless of the type of valve installed, the seats should be blessed with a three-angle valve job. As opposed to a 45-deg. valve seat meeting a 70-deg. machined surface (as is the case with a stock valve job) a three-angle valve job includes a 45-deg. seat, 60-deg. intermediate cut and the 70-deg. bottom cut. A 30-deg. machined surface is also included to blend the other side of the valve seat into the combustion chamber floor. A three-angle valve job can easily add 10-15 hp.

If not already so equipped, a V6 engine can benefit from installation of 1.72" intake and 1.42" exhaust valves, which are available through Chevrolet dealers and a few aftermarket suppliers such as Speed-Pro. For either V6 or V8 applications, the valve itself can often be modified to improve airflow passing around it. Most engine builders add a back cut (varying between 30 and 45 degrees) to eliminate sharp edges on the back face. This increases airflow even further.

In most performance applications, the factory iron valve guides are replaced with lower-friction bronze guides or guide liners. This modification isn't necessary for most street applications but is a definite advantage for super high performance or

Rocker Arms

Standard Chevrolet small-block rocker arms have a theoretical 1.5:1

Roller camshafts need some method of keeping the lifter in proper alignment. This lifter guide retainer (part no. 14101116) is used on 1987-and-later V8 blocks using an original equipment roller camshaft. It will not work with blocks not originally designed for hydraulic rollers. Aftermarket lifters must be installed instead.

ratio; in practice, very few stock rockers meet this specification, so actual valve lift is less than the theoretical specification. Consequently, a relatively painless way of increasing performance is to install accurate 1.5:1 or higher ratio 1.6:1 rockers. As an example, Competition Cams offers Magnum roller tip rockers in either 1.52:1 or 1.6:1 ratios. These relatively inexpensive rockers have proven very popular because of their low cost and excellent durability.

High ratio rockers rarely work miracles, however. Chances are, you won't experience much of a performance improvement after installing a set with a 1.52:1 ratio. A bracket racer, who has quarter-mile performance dialed in to the hundredth of a second, might be able to document a measurable improvement, but on the street the difference won't be enough to feel.

With 1.6:1 rockers, the situation can change somewhat. High ratio rockers provide the most impressive results when mated to a stock cam; improvements can range from 5 to 15 horsepower. Power increases are typically much less dramatic when a high performance cam is calling the shots, because compared to a stock camshaft, lift and valve opening and closing rates have already been increased. In most cases, with a high performance cam, the valve is already out-flowing the port, so lifting it higher and moving it at higher velocities doesn't pay much of a dividend. So if you're keeping the stock cam, you may want to consider installing 1.6:1 rockers.

CAMSHAFTS

Late-model Camaro V8s use two basic types of camshafts. On '82 through '86 V8 models, a hydraulic flat-tappet camshaft served as original equipment. Beginning in 1987, hydraulic roller lifters, and the cams required to actuate them, became standard equipment. As opposed to the link bars found on aftermarket roller tappets, the stock Chevrolet method of roller lifter retention employs a sheet metal "spider" that is bolted into the block's lifter valley. The spider holds eight retaining bars in place on top of the lifter bores. With roller lifters, some method of retention is required to prevent the lifters from spinning.

Selection Guidelines

In order to select the best camshaft for any particular application, it's necessary to understand a little bit about some of the terminology used in camshaft design. Cam profiles vary with aftermarket manufacturers as to their design and where in the rpm scale they deliver power.

Duration—To simplify things a little, the duration of the camshaft will determine where in the rpm range

The alignment bars that prevent stock roller lifters from spinning in their bores are held in place by a sheet metal "spider" that's held to the block by three bolts. This arrangement has proven to be both cheap and effective. However, it can only be installed in late-model blocks that were built to accept hydraulic roller lifters.

As horsepower increases, so does the need for heavy-duty equipment. This engine is fitted with Air Flow Research cylinder heads sporting screw-in studs and pushrod guide plates. Note 12-point head bolts.

the power will develop. If duration is increased, the power band moves up the rpm scale. Shortening duration moves the power down to lower rpm levels.

Since the duration of a cam can be measured from any starting point on the cam's lobe, many manufacturers have agreed that a point of reference should be established. Thus, you will see a cam's duration measured at .050" lift. Using *duration at .050"* lift establishes a level playing field on which cams from different manufacturers can be compared.

Lift—Increasing the valve lift generated by a camshaft allows an engine to produce more power throughout the whole rpm range. Thus, higher lifts will increase power without affecting low-end performance. In theory, the best cam is one with a relatively short duration for low end power and a high lift to create more overall power. Unfortunately this ONLY works in theory and it's physically impossible for a short duration camshaft lobe to lift a valve very high. However, it should be noted that roller cams usually offer higher lifts than a flat-tappet cam of the same duration.

Cam Facts & Figures

Over the years, Chevrolet has made several changes in the profile of the camshafts installed in 305 and 350 Camaro engines. The 1986 TPI engines are equipped with a hydraulic flat-tappet camshaft (part no. 14094097) that offers only 178 degrees of intake and 194 degrees of exhaust duration with lift of .350" and .385". These specifications are a significant step down from the cam used in 1985 TPI engines which can be found hiding under part number 14088843. That cam features intake and exhaust durations of 202 and 206 degrees with .403" intake and .415" exhaust lift and a 114.5 degree lobe separation.

A roller lifter version of this cam was installed in 1987 350 TPI engines and 305 TPI engines with manual transmissions. Listed as part number 14093643, it has 202/207 intake/exhaust duration, 404"/.415" intake/exhaust lift and 114.5-degree lobe separation. The cam used in 305 TPI engines mated to automatic transmissions was a hydraulic roller version of the earlier "peanut cam" and has only 179/194-degree intake/exhaust duration, .350"/.384" intake/ exhaust lift and 109-degree lobe separation. It's listed as part number 10088155.

The 1988-89 305 and 350 TPI engines are equipped with the same cam (part number 10066049) which offers 207/213-degree intake/ exhaust duration and .415"/.430"

As might be expected, there are several excellent flat-tappet hydraulic cam profiles available from specialty cam grinders. Many, such as the Pure Energy grinds from Competition Cams are 50-state legal.

lift and 117-degree lobe separation. For 1990 through 1992, these engines were fitted with part number 10111773 which is a slightly milder grind. It offers 202/207-degree intake/exhaust duration, .413"/.428" lift and 114.5-degree lobe separation.

Carbureted Engines—Carbureted Camaro engines have also been treated to a variety of camshafts. In spite of there being three different part numbers, specifications are virtually identical. Part number 14014475, which was used in the 1982 LG4 has intake and exhaust durations (at .050" lift) of 177 and 194 degrees, .357"/.390" lift and 109-degree lobe separation. Later versions of this cam, part numbers 14060653 and 14088841 have the same duration, but slightly less lift, raising the intake valve only .350" and the exhaust valve .385". In spite of the fact that these two cams have identical specs, the 14088841 cam is physically different because its lobes are wider.

Although hydraulic roller camshafts are considerably more expensive than their flat-tappet counterparts, the cost of a cam swap isn't outrageous because the original lifters can be reused. Most high performance camshaft manufacturers offer a variety of hydraulic roller camshafts designed specifically for late-model engines with factory original roller cams. Retrofit kits are also available to install hydraulic roller cams in 1986-and-earlier engines. These tend to be rather expensive because lifters, pushrods, valve springs and hardware are generally included.

All Camaro 2.8 and 3.1 liter V6 engines are equipped with the same camshaft. Sold as part number 14031378, it has .394" intake and .410" exhaust lift with 196 degree intake and 202 degree exhaust duration (at .050" lift) with a lobe separation of 109 degrees.

Recommendations

For a late-model Camaro V8 with a computer-controlled engine, a cam should be selected with duration at .050" lift of 202-205 intake and 206-214 exhaust. Cams with these specifications will typically have .415"-.460" lift. Many of the factory and high performance camshafts with these specifications are emissions-legal and are computer compatible.

As an example, Competition Cams offers a "Pure Energy" flat-tappet hydraulic camshaft for 1987-and-earlier carbureted small block engines that is emissions-legal in all 50 states. This cam has 202 degrees of intake and 212 degrees of exhaust duration (at .050" lift) and .429" intake and .438" exhaust lift. It's ground with 110 degrees of lobe separation so it produces strong mid-range torque.

Another option is the hydraulic roller ZZ3 cam that's part of Chevrolet's H.O. engine package. With duration numbers of 208 intake and 221 exhaust at .050" lift and a .474" intake and .501" exhaust lift, this cam offers great mid-range power with a stock type idle and is available from the local GM dealer as part number 10185071. Unfortunately, although it will probably slip by a tailpipe "sniffer," this cam is emissions-legal

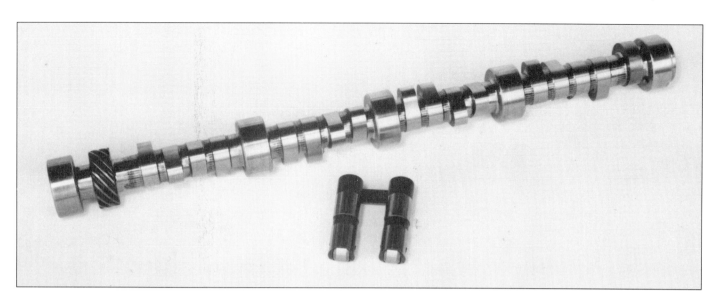

If you want to install a hydraulic roller cam in a block originally built for flat-tappet hydraulics, you'll have to use an aftermarket cam and lifters. The link bars are typically attached directly to a pair of lifters.

only as part of the complete ZZ3 engine package.

TPI Specialties offers a cam that's designed along the same lines as the ZZ3 Chevrolet grind. Designated ZZ9, this cam trades off some idle quality for increased mid-range and top-end power. Although the ZZ9 cam is designed for late-model computer controlled engines, as this is written, it is not yet emissions-legal. However, the company is committed to expanding its line of emissions-legal products, so that situation may have changed.

Obviously, any camshaft designed for a small block or 60-degree V6 Chevy engine is a potential candidate for use in a Camaro powerplant. However, a relatively mild grind is required to both meet emissions regulations and maintain computer compatibility. That being the case, it's best to select a cam that's designed specifically for late-model engines. Although this selection won't allow you to be the horsepower eliminator in a bench racing session, it will provide your Camaro with solid, usable power. And that's just what you need to build a killer street machine that can gobble up Mustangs at will.

With all the cams that are available for the small-block Chevy, picking the right one can be a daunting proposition. With third-generation Camaro engines, the best bet is to stay on the conservative side and pick a cam that offers strong bottom-end and mid-range performance.

You don't need a race car to enjoy the benefits of supercharging. Even though 6-71 blowers aren't emmissions-legal, other supercharging systems that are do an excellent job of punching up the power of third-generation Camaros.

FORCED INDUCTION 7

A forced induction system, whether it's supercharging, turbocharging or nitrous oxide injection, all but guarantees great performance and show-winning appeal when installed on a properly prepared engine. Like most small-block Chevys, late-model Camaro engines respond extremely well to all forms of forced induction. The key to success lies in the use of properly coordinated components and a bit of intelligent engineering.

When a naturally aspirated engine takes air and fuel into the combustion chamber, it does so by virtue of the suction or vacuum that's generated by downward movement of pistons in their respective cylinders. In terms of filling the cylinders with an air/fuel charge, this isn't particularly efficient. At most operating speeds, an engine is hard-pressed to ever achieve more than 85-90% volumetric efficiency (VE) because the restrictions imposed by the intake tract limit flow potential.

Paxton is one of the companies offering emissions-legal supercharging systems for TPI-equipped Camaro engines. Paxton superchargers increase output by 75-125 horsepower depending upon other engine modifications and drive ratio.

point, the engine is said to be in a "boost" condition. Nitrous oxide injection does not rely on positive pressure for its effectiveness; nitrous oxide and fuel are injected into the manifold and are drawn into the cylinders through normal vacuum. High doses of nitrous will reduce manifold vacuum which is one of the reasons that jetting of nitrous systems is critical.

When selecting any type of forced induction, make sure it's compatible with the engine as it is, or as it will be after other modifications. In most cases, the late-model V8 and V6 engines found in '82-and-later Camaros can take the extra pressures of mild supercharging, turbocharging and nitrous oxide. But if excessive boost pressures, or overly enthusiastic nitrous jetting come into play, you're sure to find at least one of your engine's weak points, and may end up with a ticket for littering when the crank and rods are scattered all over the pavement.

Tuned intake runners, such as found in TPI systems, increase volumetric efficiency by taking advantage of the inertia generated by a moving column of air. However, tuned intake systems are typically effective only within a limited rpm range; while they can boost volumetric efficiency above 100% at certain engine speeds, they also tend to reduce VE when rpm moves outside the range in which the system is designed to operate.

Forced induction overcomes normal intake tract flow limitations by force-feeding air and fuel into an engine. When this occurs with any type of supercharging, a positive pressure, rather than a vacuum exists in the intake manifold. At this

Carbureted engines can benefit from a low-profile supercharger such as this one from B&M. This system makes about 5-7 pounds of boost and can add as much as 100 horsepower, depending on the application.

Centrifugal-type superchargers, such as those produced by Paxton and Vortech are similar in construction to a turbo. The most obvious difference is in the drive section; a turbo is spun by exhaust gases while a centrifugal blower is belt-driven from the crankshaft. Since the blower is always spinning at a percentage of crankshaft speed, lag is eliminated.

SUPERCHARGING

One of the most effective methods of forced induction is supercharging, which has been around for quite some time. Depending on the ratio at which the supercharger is driven (relative to engine speed) power increases can be anywhere between moderate and mind-bending. Selecting the proper supercharger for your application depends on your budget and desired horsepower increase. For example, if you're building an all-out Pro Street vehicle, you're obviously going to include serious engine modifications—items such as low-compression pistons, high performance heads, forged crankshaft, and rods and high lift camshaft. Although this type of an engine can deliver as much as 650 horsepower, it may not serve very well as a daily driver. And chances are, it won't be greeted very enthusiastically by your local smog check station.

Supercharger Types

As far as the late-model V6 and V8 engines are concerned, there are several supercharger manufacturers who specialize in systems that are designed for unmodified, stock engines. Some of these manufacturers, such as Paxton and Vortech, have developed systems that bolt directly to your late-model small block, whether it is carbureted, throttle body injection or a TPI system. Many of these systems are also emissions-legal.

The limited amount of boost produced by these systems is ideal for a stock engine with cast pistons, cast-iron crankshaft and unported cylinder heads. Although power increases will be limited compared to a Roots-type "Jimmy" blower, they're right in line with the additional power levels a stock engine can handle. When installed on a 350 cid Camaro engine, a street-legal supercharger can mean going from 245 horsepower to approximately 350 horsepower.

Another great benefit of bolting on a supercharger is if you should later decide to add items such as a camshaft, headers, larger manifold or TPI runners, the supercharger will respond with additional horsepower. Most superchargers designed for stock Chevy engines will deliver anywhere from around 5-7 psi of boost, which is the practical limit for a stock engine. Higher boost levels may result in serious problems, and will create the need for fuel octane levels above those available at the gas pump.

Roots Type—There are two basic types of superchargers that have become popular for use on engines in street-driven vehicles. 1982-and-later Camaros equipped with a carburetor can accept either a Roots or centrifugal supercharger. The Roots type offers the most visual appeal, and some low-profile models will fit beneath the hood of a late-model Camaro.

A Roots supercharger bolts to a special intake manifold and includes a mounting pad or adapter which accepts the carburetor.

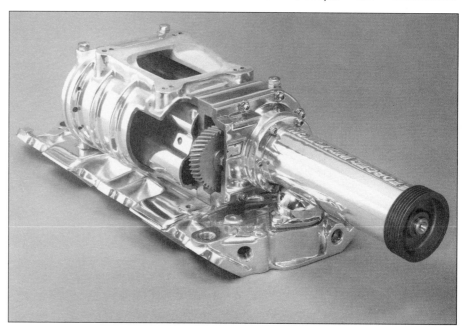

The Roots-type supercharger employs rotors that capture air between their lobes, compress it and force it into the engine. Roots blowers set up for street engines draw air in through a carburetor or fuel injection throttle body and typically produce 7-8 pounds of boost. Boost pressures can be increased or decreased by changing drive pulley ratio.

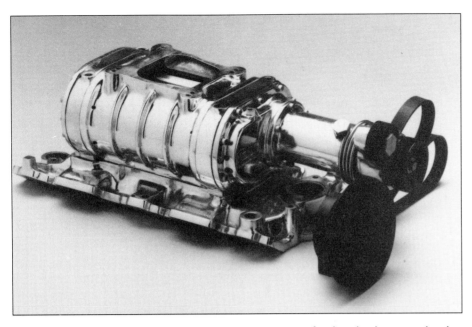

Weiand offers a number of Roots-type superchargers which accept a four-barrel carburetor or throttle body. These blowers are designed to produce up to 7 or 8 pounds of boost.

Manufacturers such as B&M and Weiand manufacture superchargers suitable for a stock, late-model carbureted V6 or V8 engines. Both manufacturers have created kits for various engines, so installation is rather straightforward. In most instances, rejetting the existing carburetor or installing a new, larger one is necessary to achieve optimum performance.

Centrifugal—For late-model Camaros with Tuned Port Injection, a Roots supercharger can't be installed (unless the TPI system is discarded). But centrifugal superchargers work extremely well with TPI engines. Since these superchargers blow through the TPI's throttle body assembly, the ECM can handle its engine management chores if a special PROM is installed.

Centrifugal superchargers are similar to turbochargers in that they use a small turbine with an air intake on one side of a housing and "blow" air out the exit on the other side. The difference is that a centrifugal supercharger is driven by a belt attached to the crankshaft pulley, whereas a turbo is powered by hot exhaust gases. As a result, a centrifugal blower delivers instant power when you step on the throttle as opposed to the slight lag that characterizes turbo operation.

Since the impeller of a centrifugal supercharger is always spinning, the need for proper lubrication and heat dissipation is very important. Thus many centrifugal superchargers use some sort of built-in oil system for this purpose. The Paxton supercharger (which has been around for over 40 years, and was even used in original equipment installations dating back to the late Fifties) relies on F-type transmission fluid (which is held in a self-contained reservoir) to cool and lubricate itself. The Paxton units are ideal for installation on TPI engines and they are also emissions-legal.

Similar to the Paxton is the Vortech supercharger, which is also a centrifugal design and operates off a series of internal gears. Both units provide a Camaro tuned port injected engine with anywhere from 5-7 psi of boost to deliver as much 100 extra horsepower from a stock V8.

Complimentary Components

When any kind of supercharger is installed, there are obviously some other changes that should also be made. For example, a computer chip change is usually in order to tailor

It's a tight fit, but a centrifugal supercharger will fit beneath the hood of a third-generation Camaro. This Paxton model blows through a modified TPI system.

When adding a supercharger to a TPI engine, fuel injector flow capacity may need to be increased. Autotronic Controls (makers of MSD ignitions) has an extensive line of fuel injection components, including injectors with flow rates of up to 50 pounds per hour. As a comparison, stock 305 injectors are rated at 19 pounds per hour.

air/fuel ratio and ignition timing to the requirements of an engine operating under boost. To further enhance the power output of a supercharged engine, a high performance ignition should be added. With higher than normal pressures being present in the combustion chambers, spark intensity should be increased over the stock level to ensure proper combination.

Another excellent option is to add a control unit that retards timing when boost increases. MSD and other companies offer units that automatically retard timing according to the amount of boost present. Also available are control boxes that allow timing to be controlled manually by the driver. These help control detonation and make tuning easier.

Another unit that works well is a system from J&S Electronics. Their unit automatically retards ignition timing similar to that of the factory computer. Using a detonation sensor mounted on the intake manifold, the J&S unit "listens" for detonation and retards timing in one or various cylinders until the detonation has stopped. This unit can also be adjusted for sensitivity and the amount of timing retard. Once set, this unit rarely needs to be adjusted.

TURBOCHARGING

Turbochargers utilize an engine's exhaust gases to spin a turbine that shares the same shaft as another turbine positioned in the intake section of the turbo housing. Hot exhaust gases spin the exhaust turbine, which in turn spins the one located in the intake tract. Unfortunately, the intake turbine doesn't spin fast enough to generate significant boost until exhaust gas volume increases to a specific level. This results in the infamous turbo lag.

Turbochargers have been used for decades on small displacement four- and six-cylinder engines, to give them the power of bigger V8's. Similarly, mild boost levels in a V6 or V8 can greatly increase power output, especially on a stock, late-model, Chevy small block. However, the amount of power and boost greatly depends on the quality of the turbo system and its installation.

The advantage of turbochargers is that they deliver most of their power free of charge. With the turbo being driven by exhaust gases, no power is consumed in generating boost—although there is some loss in exhaust flow efficiency compared to free-flowing headers. Losses can be minimized with a well-designed system, but they are never completely eliminated.

HOW TO INSTALL A PAXTON SUPERCHARGER

Paxton Superchargers has come up with a street-legal supercharger kit that bolts under the hood on late-model Camaros, Corvettes, and all other GM products that use Chevy small blocks (such as trucks). The installation is simple, requiring few tools and about five hours of time, depending on how proficient you are with tools. The supercharger offers 5-7 lbs. of boost, increasing both horsepower and torque by as much as 45% with no engine modifications. Furthermore, Paxton was able to get a CARB (California Air Resources Board) exemption certification for the kit for use on all fuel-injected engines, which means it meets all required air quality standards as determined by the Clean Air Act and is smog-legal for use in all 50 states. It's an efficient means of increasing the horsepower of your stock TPI or TBI Camaro, Corvette or Chevy pickup. For more information contact, Paxton Products, 929 Olympic Boulevard, Santa Monica, CA 90404-3795. 310/450-4800.

1. The first part of the supercharger installation involves removing the air cleaner assembly, serpentine belt, alternator and power steering pump, which have all been removed in this photo. For details pertaining to your vehicle, see the appropriate Chilton's or Haynes manual.

2. The Paxton kit includes a pulley that attaches to the damper and this bracket which bolts directly to the block.

3. With the bracket bolted in place, there is enough room to mount the Paxton supercharger. The alternator is mounted below the power steering pump.

4. Since the alternator is now located below the steering pump, the Paxton kit includes extra wire for the alternator's battery hook-up.

5. With the wiring done, the Paxton unit bolts directly on to the supplied bracket.

6. The three belts are hooked up to the supercharger and the hoses are then attached. All the components for this TPI installation are chromed or polished aluminum.

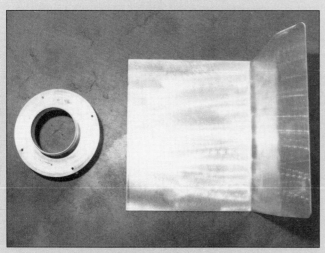

7. This box and adaptor is used to relocate the air inlet beside the cruise control valve and through the inner fender.

8. With the air box installed, cool air is fed through the K&N air filter, through the fender and routed via a hose to the supercharger inlet.

Turbo systems for late-model Camaros can also deliver impressive power increases. This twin-turbo system from Gale Banks Engineering is made for carbureted small blocks and can produce as much as 600 horsepower a properly modified engine.

Disadvantages

Understandably, being exhaust gas driven, turbos generate considerably more heat than mechanically driven superchargers. Heating of the intake mixture (which reduces power) can be minimized through use of an intercooler (if there's enough room).

Lag is always a consideration with a turbocharger. Some turbo kit manufacturers have eliminated any substantial lag with the use of centrifugal type turbos or using twin turbos for extra boost, resulting in higher horsepower levels.

Another drawback to turbo systems is that very few are designed for "do-it-yourself" installation. The exceptions are custom systems designed for specific models of vehicles. "Universal" kits typically require more time and engineering than most people are capable or willing to perform. Another consideration is that there are currently no

Nitrous oxide systems are available for both fuel injected and carbureted engines. This direct port system incorporates "Fogger" nozzles which are plumbed into each intake runner.

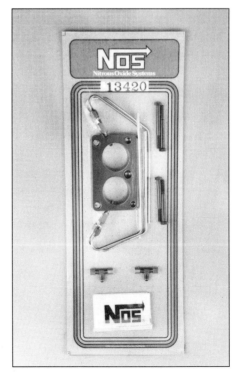

Plate systems are also available for TPI engines. Installation is relatively easy as the injector plate is simply bolted in place between the throttle body and plenum.

A nitrous oxide system can be proudly displayed or hidden. The owner of this 1987 Camaro chose to hide the nitrous tank in the compartment formerly occupied by the spare tire.

emissions-legal turbo systems for third-generation Camaros.

NITROUS OXIDE INJECTION

By far one of the most inexpensive horsepower-per-dollar improvements available for an internal combustion engine is a nitrous oxide injection system. Most kits range from about $250 to $400 for a basic 100+ horsepower unit. Nitrous oxide injection can be easily installed on any type of engine and most systems can be easily removed when a car is sold.

Although it doesn't pressurize the intake tract, a nitrous oxide system can be thought of as a chemical, as opposed to mechanical, "supercharger." When a nitrous oxide system is activated, highly compressed liquid nitrous oxide is force-fed through a solenoid valve and discharge nozzle into the intake manifold. Once past the solenoid, nitrous is no longer subjected to bottle pressure (extremely high pressure is required to keep nitrous in its liquid state) and with high pressure no longer in force, the liquid becomes a gas. Ideally, the liquid-to-gas transformation should take place at the nozzle so that the cooling effect of vaporizing nitrous liquid is maximized within the intake system.

Nitrous oxide is a chemical compound composed of two nitrogen atoms and one oxygen atom (N_2O) and by weight, it contains 36% oxygen. Nitrous oxide won't burn by itself, but what it does bring to an internal combustion party is a lot of oxygen, which is an essential ingredient in the combustion process. Also known as "laughing gas," nitrous oxide has been used as an anesthetic for quite some time.

Although there are many nitrous oxide kit manufacturers, the leader in the nitrous oxide industry is Nitrous Oxide Systems (NOS) of Cypress, CA. NOS offers a wide variety of systems designed specifically for late-model Camaro engines, including those with Tuned

Plate-type nitrous systems are available to fit intake manifolds with either a standard Holley or QuadraJet carburetor flange. These systems are compatible with electronic engine controls.

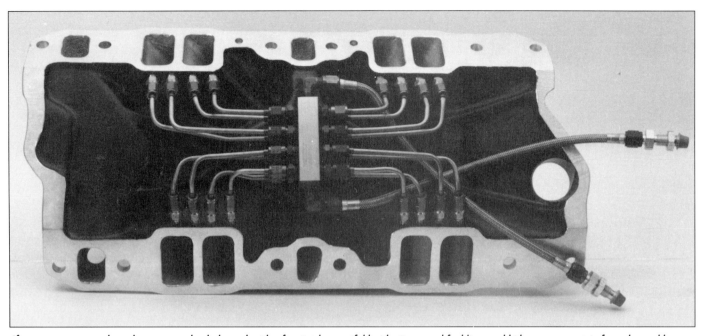

If you want to get real sneaky, you can plumb the underside of an intake manifold with nitrous and fuel lines and hide your power trip from the world. Photo by Dave Emanuel.

Port Injection. Furthermore, some of their systems are emissions-legal and others are currently being tested for certification.

Installation

Installing a nitrous system is relatively simple. Engines equipped with a carburetor or throttle body injection are most commonly fitted with a plate system. This type of system is centered around an injector plate, which contains nitrous and fuel nozzles. The plate mounts between the carburetor (or throttle body) and intake manifold. Higher horsepower systems use direct port injection with nitrous and fuel nozzles installed directly in each runner of the intake manifold.

Tuned Port Injection engines can be equipped with either a plate or direct port nitrous system. A more sophisticated system developed by NOS uses only one solenoid to inject the nitrous oxide in front of the throttle body of a tuned port engine. Upon activation, the engine's stock PROM switches over to a preprogrammed nitrous chip which uses the factory fuel injection to provide additional fuel. In recent testing of this system by NOS on a stock, late-model 350 TPI engine, up to 400 horsepower was achieved with stock injectors and a properly programmed computer chip.

The power potential with nitrous oxide is very impressive and a computerized system eliminates the need for push-button activation. It also can be programmed to turn the system on or off at various rpm levels, depending on the engine and its use. Many Camaro owners are satisfied with a simple nitrous oxide system that can provide power on demand while not affecting emissions or fuel economy during normal street driving. Besides, it's a good way to beat the pants off of a Mustang!

Any time forced induction is added to a TPI engine, it's a good idea to add an adjustable fuel pressure regulator. Increased fuel pressure helps maintain proper air/fuel ratio at wide open throttle. The computer compensates for the added fuel flow during part throttle operation so fuel economy doesn't suffer. Photo by Dave Emanuel.

EMISSIONS-LEGAL NITROUS OXIDE INSTALLATION

One of the best ways to increase your Camaro's performance at a reasonable cost, is to install a nitrous oxide system available from NOS. Their dual prom system is very simple to install and within a matter of a few hours, your Camaro can have as much as 400 horsepower on demand! This particular system is specifically for TPI-equipped models with the speed density system. There are however, similar kits available for MAS system Camaros and those equipped with a four-barrel carburetor that yield similar results. For more information, contact: Nitrous Oxide Systems, 5930 Lakeshore Dr., Cypress, CA 90630. (714) 821-0592

1. The first step in installing the NOS emissions-legal system is to route the hose and heater wires to the rear hatch area.

2. A small plug in the hatch area can be pushed out and the line for the nitrous can be slipped through and routed towards the engine compartment. The heater lines on the opposite side can be tucked under the carpet to a heater toggle switch that can be mounted in or under the dash.

3. The nitrous line can follow the factory fuel lines and should be properly secured with wire ties.

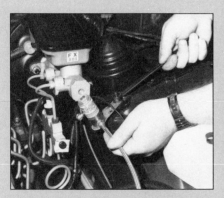

4. The nitrous line is then attached to the single solenoid which can be mounted to one of the shock tower bolts. Another line is then attached to the solenoid and is routed to the air intake duct.

5. A small hole is drilled in the air intake duct and the nitrous nozzle is inserted and bolted in place. Mounting the nozzle closer will make the nitrous "hit" harder so this area is a good place to start.

6. The dual E-PROM unit is attached to the stock computer which is located under the dash in the passenger side compartment.

7. With the stock computer out, the cover plate is removed and the factory chip is unplugged from the computer.

8. The NOS E-PROM switch box is then plugged into the factory location for the PROM chip. With the NOS box plugged in the cover plate is not reinstalled.

9. The cover on the NOS box is removed and the factory chip is plugged into the box. This will allow the use of the factory chip during normal operation and will automatically switch to the NOS chip during wide open throttle. The NOS chip will tell the computer to provide extra fuel to add to the nitrous injection.

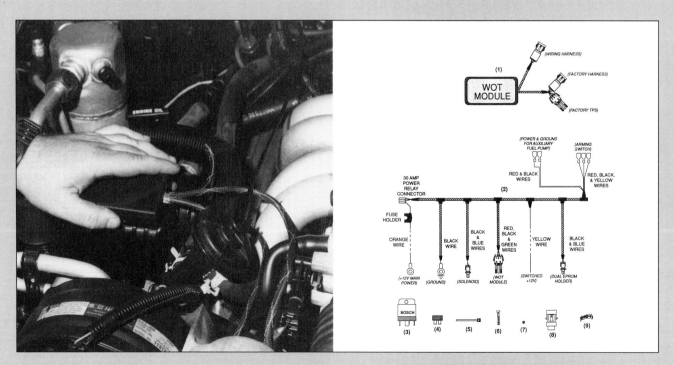
10. The NOS control box is mounted in the engine compartment away from heat. A good area is on the air pump diverter valve.

11. The wiring harness from the control box is connected to the throttle position sensor on the throttle body. The factory harness is then plugged into a female end of the harness.

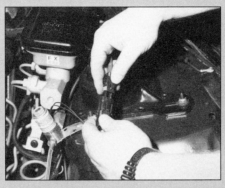
12. The harness also plugs into the NOS solenoid.

13. The 20-lb. bottle can be mounted on this trick board made to fit in the Camaro's trunk area. This board is not included in the kit but can be easily made from particle board with some careful measurements.

14. Once the bottle brackets are bolted to the board, the bottle is inserted and the entire assembly is positioned into place.

15. The nitrous line is then connected to the bottle valve. At this time, all the lines should be double-checked. There should be no leaks in the system.

16. The NOS kit also comes with this executive order number sticker that indicates the system is emissions-legal. You'll want to add this to satisfy smog-check officials.

17. This installation features a custom-made switch panel that uses the ash tray area. Here, the optional pressure gauge was installed along with the heater switch and arming switch on the far right.

18. With everything hooked up, the factory 5.7-liter TPI V8 can produce as much as 400 horsepower! That's almost double the stock horsepower output and enough to propel your Camaro to 12 second quarter mile blasts.

Attention-grabbing, neck-snapping acceleration is the result of a strong engine, good traction and a solid drivetrain with the proper gearing. Photo courtesy of All Chevy Magazine.

DRIVETRAIN

As the saying goes, a chain is only as strong as its weakest link. When it comes to late-model Camaros, the weakest link in most models, is the drivetrain. From '82-and-up, Camaros came equipped with a variety of transmissions and rear ends that for the most part, leave much to be desired in the performance category.

It is not difficult, however, to change or modify the existing drivetrain to be able to handle the extra loads and torque of a 350 cid or larger high performance engine. There is an abundance of factory and aftermarket transmissions and rear axles that are strong enough for the most demanding drivers.

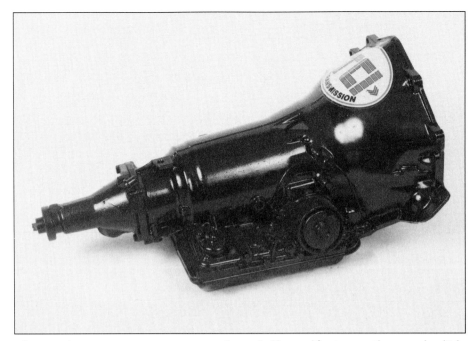

If you're reluctant to tear your automatic apart for a rebuild or modification, consider a complete high performance transmission. With all the latest heavy-duty parts, you'll get improved performance and durability.

Equipping your Camaro with a high performance transmission and rear end is something you should definitely think about if you've modified your engine to produce anywhere over 350 horsepower and 375 lbs-ft. of torque. This is especially a consideration if you plan to do any serious racing or have aspirations of running quicker than 13-second quarter-mile times.

AUTOMATIC TRANSMISSIONS

Depending on the year and model of your Camaro it was originally equipped with either a Turbo Hydra-matic 200 ('82-'83 models) and the Turbo Hydra-matic 700-R4 ('83-up). At first, enthusiasts disliked these transmissions and often they were replaced with the more dependable and durable Turbo Hydra-matic 350 or 400 automatics. The TH-200 is not particularly suited to high performance applications. There are various aftermarket kits available to swap in a TH-350 or TH-400 in place of the TH-200. All you'll need are a few brackets, which are available from B&M and Gale Banks Engineering. The best bet however, would be to swap the TH-200 for a 700-R4. There are no brackets required for this swap and all you need to do it is to use the 700-R4 components such as the driveshaft, shifter and mounts that are original equipment on the 1983-and-later Camaros that had 700-R4 transmissions factory installed.

The 700-R4

The Turbo Hydra-matic 700-R4 transmission debuted in the 1982 Corvettes and has since seen duty in just about every rear-wheel-drive GM vehicle, from Camaros and Firebirds to full-size pickups and S-10 Blazers. As successor to GM's 3-speed automatics, the 700-R4 offered the advantage of a fourth, or overdrive gear, and promised much in the way of fuel economy and driving comfort. However, it would quickly become apparent that the 700-R4 was not, to put it mildly, without some serious flaws. As in most cases of automotive development, Hydra-matic engineers were faced with striking the best compromise between performance and economy. Because the 700-R4 would be employed in a variety of vehicles, it was designed to perform under a wide variety of driving conditions while achieving the highest

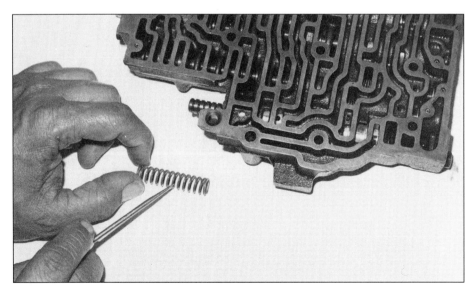

The 700-R4's valve body can be modified and recalibrated to provide crisper, firmer shifts. Companies like TCI and Art Carr offer kits to accomplish this, or you can ship your tranny to them and they'll handle it for you. Photo by Michael Lutfy.

possible fuel economy, which is a lot to ask of a single transmission.

Because of its broad-based design, the first thing most enthusiasts did with a 700-R4 transmission was to get rid of it by swapping in a suitable replacement. But aftermarket items have enhanced the durability and performance of the 700-R4 (the factory has also steadily improved the transmission as well) and it has quickly become the performance transmission of choice. The 700-R4 has the added advantage of a fourth gear overdrive and a lower first gear ratio. The first-through-fourth gear selections are 3.06:1, 1.62:1, 1:1 and .70:1. This combination enables the 700-R4 to deliver strong performance and excellent fuel economy.

Swapping out a 700-R4 automatic for a TH-350 or 400 would not be a wise choice, unless you are building an all out drag car, that might see some 10-second or faster quarter-mile times. But even then, there are some 10-second quarter-mile Camaros and Corvettes that use a modified 700-R4 quite successfully.

As mentioned there are various components available to improve the performance and durability of 700-R4 transmissions. One of the first items to consider is a high performance servo assembly. This will increase the 700-R4's line pressure and give your Camaro quicker shifts that love to chirp tires. The servo assembly can be purchased through your local Chevrolet dealer and includes the following parts:

Spring pressure regulator	#8639164
Boost valve	#8634940
Servo piston	#8642079
Servo housing	#8642110

High performance transmission parts are also available from companies like Art Carr (Fountain Valley, CA) and TCI Automotive (Ashland, MS). In addition to complete, ready-to-run transmissions, both companies offer everything from high output 10-vane pumps to modified valve bodies, shift improving kits, heavy-duty servos and high performance bands and clutches.

TCI recommends that if you want to build a trans that will stand up behind a healthy engine, the best bet is to start with a 1987-or-later core. The 700-R4 has had a problem with the 3-4 clutch pack since day one. The 1987-and-later models have a revised input drum that solves most of these problems. It's not feasible to retrofit the late drum to earlier transmissions. It is also recommended that you use commercial-type 3-4 clutches, which are thinner (.064" instead of .080") than the standard ones. That allows you to load the drum with 8 or 9 clutches instead of 5 or 6, which you usually find. Another advantage of the 1987-and-later transmissions is that the pump body was revised to include a stepped bushing bore, which prevents the bushing from moving forward into the front seal, which always results in severe fluid leakage."

Art Carr's approach centers around the torque converter and the flow of oil through the system. He essentially eliminates the factory-designed restrictions by installing a non-lockup torque converter, and

Stock automatic transmissions aren't known for snappy shifts. That situation can be corrected with a transmission reprogramming kit like this "Trans-Scat" kit from TCI Automotive.

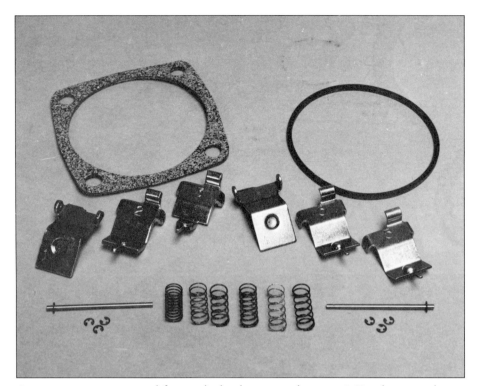

Some automatic transmissions shift too early, thereby putting a damper on 0-60 and quarter-mile times. TCI Automotive offers this governor spring and weight kit which allows shift points to be adjusted as necessary. Photo by Dave Emanuel.

reprogramming the transmission to increase the volume and pressure to full flow the transmission with more oil at higher engine speeds.

Shift Improving Kits—Stock transmissions are built with an eye towards pleasing many masters, some of whom have sensitive derrieres that get excited if a transmission shifts too firmly. So ever since high performance maniacs have been using automatic transmissions, the need for crisper, firmer shifts has existed. Shift kits meet that need by recalibrating the valve body to provide the aforementioned crisper, firmer shifts.

Before purchasing a shift kit, however, make sure that it has the proper gaskets and that you are able to install it. Some kits require drilling of pressure relief holes in the valve body or spacer plates. If you're not willing and able to read and follow instructions, these types of modifications should be done by an experienced transmission specialist.

Shifters—Along with a good shift kit, a performance shifter can be an added benefit if you plan to shift the transmission manually. A performance shifter has special stops that make it all but impossible to select the wrong gear or to shift through a gear. Fortunately, there are a number of shifters available that fit the factory console and are relatively easy to install.

Some of the more popular shifters are those from Hurst and B&M. Both of these manufacturers make good quality, performance shifters. Hurst has a very unique shifter that has a dual gate. One gate is used for normal driving; moving the shift lever to the other gate makes it a racing shifter with a lock-out feature to prevent engaging the reverse gear.

No matter what type of shifter you choose, make sure it is compatible with your transmission and that you check for worn parts and missing clips to ensure safe and efficient performance, even from your stock shifter.

Torque Converters—Another way to improve the 700-R4 is the use of a performance torque converter. Torque converters vary in size and rpm stall speeds. The amount of stall depends on the performance of your

Convincing an automatic transmission to shift into the appropriate gear at the appropriate time can be challenging. Aftermarket shifters, like this one from Hurst, can prove helpful. It also adds a racy look to the interior.

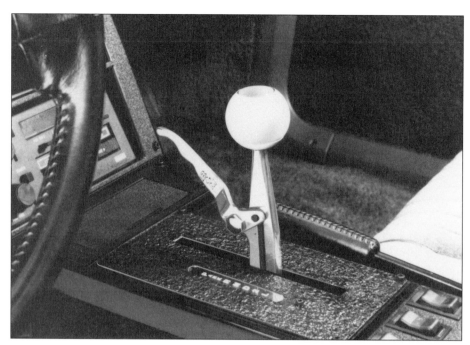

B&M also offers an automatic shifter for third-generation Camaros that fits in the original console. This feature makes for a clean installation.

engine, tire size and axle ratio. Most street and strip Camaros use anywhere from a 2,500 to 2,800 rpm stall converter. Higher performance engines making as much as 375-400 horsepower should consider a 3,000-3,800 rpm stall converter. Your best choice on a performance torque converter should come after contacting various manufacturers and discussing with them the details of your Camaro.

Although some high performance transmission manufacturers disable the torque converter clutch, there's no need to do so. Some companies offer modified stall speed torque converters that retain full lock-up capability. Even though these converters are "looser" than their stock counterparts, fuel economy isn't affected very much—if at all.

In the 700-R4, the computer controls torque converter lock-up. In most instances, the converter will lock when the transmission shifts into third gear—even at wide open throttle. It is possible to install an override switch and prevent the converter clutch from locking, but this is rarely required. It's more of an option for testing purposes.

A good torque converter should have brazed fins, and incorporate a Torrington thrust bearing for optimum performance. If you have a very high stall converter (3,000 stall or higher) you should consider an external oil cooler to aid in transmission cooling. Smaller, higher stall converters generate much more heat than stock converters do, and sometimes, the factory cooler isn't adequate. These types of converters can actually be counterproductive in a late-model Camaro. Remember, these are not high rpm engines, so if stall speed is too high, there won't be enough usable engine speed for most driving situations.

How effective is a high performance torque converter? During testing it completely changed the car's personality. When the stock converter had been in residence, nailing the accelerator from a dead stop produced mild wheel spin. With a "TCI Street Fighter" in place, the same gas pedal action absolutely blazed the tires; the car left the starting line like there was a 350 rather than a 305 under the hood. At the drag strip, times improved by over two-tenths of a second.

MANUAL TRANSMISSIONS

Although many Camaros were blessed with the 700-R4 automatic,

One of the best performance values available for Camaros with automatic transmissions is a high performance torque converter. It can really make an unbelievable difference in acceleration.

TRANNY SWAP
Installing a Richmond 5-speed transmission

Some Camaro enthusiasts love banging those gears and prefer to shift themselves rather than letting an automatic transmission do it for them. Although the automatic transmissions can be beefed up to handle over 400 horsepower, the factory Borg Warner T-5 manual transmission is not one to handle any more horsepower than what comes on the car stock.

For this reason, Camaro enthusiasts with a bit of horsepower under their hoods, should consider a tranny swap. Although installing a ZF six-speed from a Corvette sounds like a great idea, be prepared to pay a hefty price for one of them...if you can find one. Installing a six-speed is also not as easy as you might think, and requires some cutting of the floor and trans tunnel to accommodate it.

By far, the easiest transmission to swap in place of the T-5 is the Richmond 5-speed. With two models now available, a 5-speed with 1:1 fifth gear or a six-speed with a 1:1 fifth and an overdrive sixth gear, the choice is simple. The Richmond 5- or 6-speed transmissions bolt easily into the Camaro's transmission tunnel and works with factory or aftermarket clutches and only requires the purchase of a Lakewood bellhousing and performance shifter.

The shifter of choice is a Long Engineering shifter, since it has very short throws and works within the confines of the factory center console without any cutting. The Long shifter also comes with the necessary transmission mount adaptor and has a provision for the transmission torque arm mount. With the Richmond 5-speed, the drivetrain can handle as much as 550 to 600 horsepower which makes banging through the gears a whole new experience.

Sources:

Richmond Gear
Box 238
Old Norris Rd.
Liberty, SC 29657
(608) 843-8800

Centerforce
Division Of Midway Industries
7171 Patterson Dr.
Garden Grove, CA 92641
(714) 898-4477

Lakewood Industries
Division Of Mr.Gasket
8700 Brookpark Rd..
Cleveland, OH 44129
((216) 398-8300

Long Shifters
RD#2
Annville, PA 17003
(717) 867-1303

1. This Richmond 5-speed transmission is almost unbreakable and is the easiest transmission to install if you are looking for something to handle more power from your Camaro. It mounts the same as the factory transmission by using a Lakewood bellhousing.

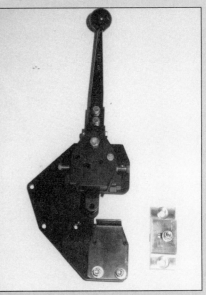

2. The secret to this installation is using the proper shifter. This shifter from Long Engineering is the only way to go. It features short positive throws and comes complete with tranny mount and adaptor for the torque arm..

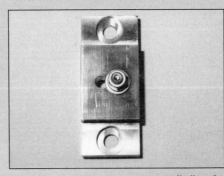

3. The Long transmission mount will allow the Richmond 5-speed to be mounted to the factory trans crossmember. The only drawback is that it is a solid mount made from billet aluminum. This will increase transmission noise and vibration in the interior of the car, but there's always some price to pay to go faster.

4. As you can see here, the trans mounts nicely to the factory crossmember. Note the small notch on the crossmember to accommodate one of the transmission bolts.

5. The special Lakewood bellhousing #15020 has the proper bolt holes for the Richmond 5-speed and makes bolting the transmission easy to the late-model engines. This bellhousing also features a proper clutch bracket so that the factory hydraulic clutch can be retained.

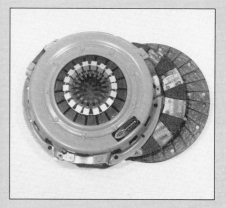
6. Although you can use a factory clutch, it is highly recommended that you use a performance dual friction clutch like this one from Centerforce. This particular clutch can handle low to high horsepower levels and is specifically made for Camaros using a hydraulic clutch piston. This makes for effort-free clutch feel with superior engagement.

7. The bracket for the Long shifter tucks in close to the transmission, allowing for free movement during shifting. It also houses the stock torque arm and makes installation of this set-up a breeze.

8. With the shifter installed, the factory torque arm bolts up easily with a factory torque arm insulator. The insulator and bracket are available from your Chevy dealer or if your old one is good, it can be reused here.

9. With the proper items, the transmission can be easily installed. In some installations, you can use a 1-1/2 inch shorter yoke to avoid shortening and rebalancing the driveshaft.

This 25th Anniversary Camaro is a Chevrolet engineering project that uses a Corvette six-speed manual transmission for increased power and improved fuel economy.

there are a good number that were cursed with a poor five-speed manual transmission. Camaros from 1982 were the only ones with either a Super T-10 or Saginaw four-speed manual. These are very durable gearboxes (especially the Super T-10) and entirely suitable for high performance applications. The only problem is that they are both four-speeds, so there's no overdrive ratio.

Tex Racing Transmissions (Ether, NC) builds a race ready Super T-10 transmission and can put one together that will fit a third-generation Camaro. Although the Saginaw trans is good for the street use, it doesn't have sufficient strength for racing. If you're going to spend any money for a bolt-in manual transmission, go for the Super T-10.

For the 1983 model year, Chevrolet dropped the previous transmissions in favor of the Borg-Warner T-5 five-speed. Two versions were manufactured, one for 1983 to '87 vehicles, the other for 1988-and-later Camaros. The early version of the T-5 was a bit too lightweight and failed regularly. It was commonly thought that the transmission just wasn't up to the task of handling any type of horsepower until someone noticed that the T-5s used in Mustangs held up considerably better. Consequently, the Camaro transmission was upgraded for the 1988 model year. The late-model T-5s are said to be able to handle up to about 350-375 lbs-ft. of torque.

Later model T-5 transmissions can be easily swapped for the earlier models, as long as the clutch linkage matches. Only the '83 T-5 transmissions were combined with mechanical clutch linkage; later models used a hydraulic setup. That fact should also turn on a light bulb—if you want to swap a T-10 or Saginaw into a 1984 or later Camaro, you'll need the appropriate early model bellhousing and linkage arrangement.

Aftermarket Choices

Serious performance Camaros making anything more than 325 honest horsepower, need an aftermarket transmission. The stock T-5 just won't handle the torque. A much better choice is a Richmond five-speed or a G-Force Lethal Weapon from G&G Specialties (Minneapolis, MN). All of these transmissions are capable of handling the power of a big block, so they should be more than up to the task of handling anything a small block, even an exceptionally healthy one, can dish out.

The Richmond Gear and G-Force transmissions are essentially the same in design, except that the Lethal Weapon is built specifically for Pro Street and race applications. It also features a special face tooth gear engagement system in place of standard synchronizers. This makes for a much more durable transmission that can be shifted with lightning speed—with or without the clutch.

The Richmond five-speed is a good street and strip unit that can handle lots of horsepower; however, it can be balky during a speed or power shift. Both the Richmond and G-Force transmissions fit easily into the Camaro's body, requiring only minor modifications to the rear mounts, and a shorter length driveshaft. These transmissions also require a standard Corvette bellhousing or an aftermarket bellhousing such as a Lakewood or McLeod.

Swapping

The Corvette six-speed looks like another attractive candidate for a transmission swap, but it's difficult to find one and these gearboxes are outrageously expensive. A much better choice would be a Borg-Warner T-56—the six-speed box that's original equipment in 1993- and-later Camaros (the fourth generation). This is an ideal street/strip transmission with excellent shifting action and adequate strength for "severe service." At this writing, there isn't information available about transmission swaps involving the T-56, but transmission specialty shops should be able to answer your questions.

Clutches

Standard Camaro clutches are fairly reliable for performance use. The stock clutch system incorporates a diaphragm-style pressure plate and 10.4" diameter clutch disc with a hub that accepts a 1-1/8" diameter, 26-spline transmission input shaft. The stock clutch is adequate for

Before the Borg-Warner T-56 six-speed (as used in 1993 Camaro Z28s) was available, some people tried installing Corvette six-speed gearboxes in Camaros. They found it to be very expensive. The T-56 is definitely a better way to go.

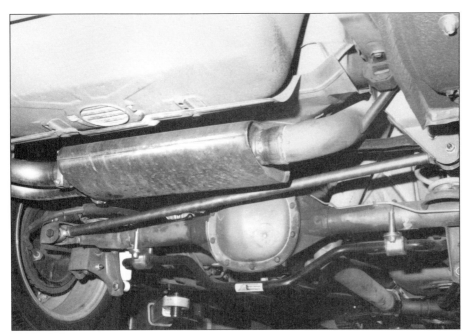

This 7-1/2-in., 10-bolt Saginaw rear was factory-installed in 1982-and-later Camaros. For performance applications, it is best to use the latest version of this rear because it offers increased strength. However, the strongest rear axle assembly is the Borg-Warner version that was used on a limited basis. It can be most easily identified by its nine-bolt carrier cover. Unfortunately, ring-and-pinion selection is non-existent.

clutch disc installed in a street-driven Camaro (or any other vehicle) has a marcel and a sprung hub.

"Marcel" is a term that applies to a wavy spring that resides between the clutch linings. It cushions clutch engagement and makes for much smoother operation. Clutch discs without a marcel are intended for racing where smooth engagement is not an important consideration.

Similarly, discs with a solid, rather than a sprung hub don't provide adequate cushioning for smooth engagement. Although it may not be possible to feel the difference during normal starts, a clutch disc with a solid hub places much higher stress on the driveline because there's no "shock absorber" to soften the hit when the clutch is engaged. This can prove disastrous with a T-5 transmission if the driver's right foot gets too enthusiastic before the left foot is moved off the floor.

REAR ENDS

The standard Camaro rear end carries a reputation of a poor quality most street/strip Camaros, the exception being those with highly modified engines and wide, sticky slicks mounted on the rear wheels.

Installation of a high performance pressure plate requires use of a different hydraulic throwout bearing, such as those offered by Tilton or McCleod. High performance clutches are available from a variety of clutch manufacturers. But regardless of the supplier, make sure that any

Another option for high performance Camaro, is a Dana 44 rear. Although no longer available from Chevrolet, it can still be found at swap meets and through some aftermarket suppliers.

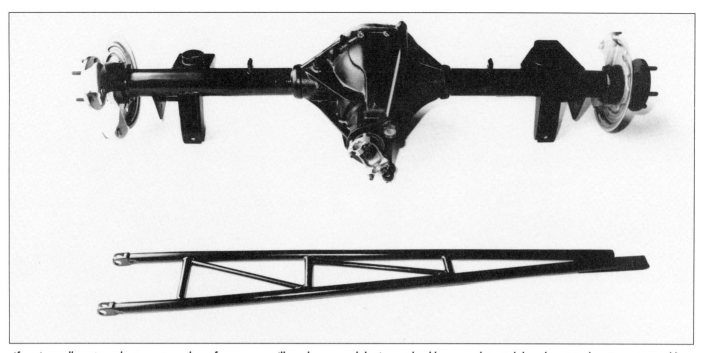

If you're really serious about quarter-mile performance, you'll need a rear end that's considerably more substantial than the original equipment assemblies. One option is a 12-bolt Chevy rear, specially modified by Summers Brothers. This rear uses C-clip eliminators and 30-spline axles and has the brackets needed to bolt into a third generation Camaro. This rear can handle 700 horsepower, slicks and 9-second quarter-mile blasts.

unit that can not withstand the rigors of high performance use. The fact is that the stock rear ends on many street and strip Camaros have endured the rigors of drag strip and endurance road racing. On the other hand, however, many of them have failed with less strenuous use.

Stock Differentials

There were three different rear ends available for 1982-and-later

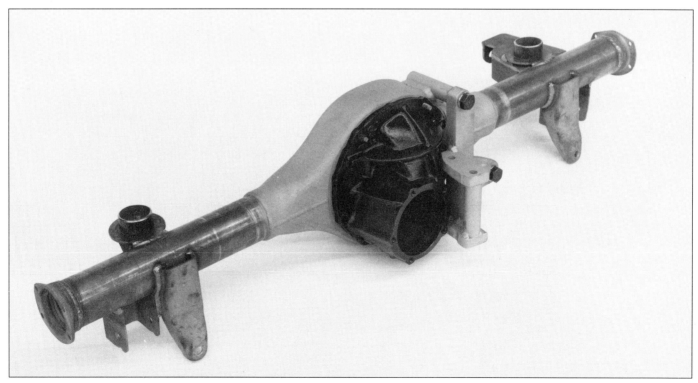

The Ford 9" axle assembly—if it's strong enough for a Pro Stocker, it can handle whatever a street-driven Camaro can dish out. Several manufacturers offer 9" Ford rears with the spring perches and brackets necessary for installation in a 1982-'92 Camaro.

Differential cases must be changed when switching from a 2- to a 3-series ratio (or vice versa). If a switch has to be made, the sensible thing to do is install an Auburn Gear limited slip differential which is stronger than the stock model. It's also cheaper because it's available through aftermarket dealers and distributors. Photo by Dave Emanuel.

Dana 44—Sold as a GM racing rear end assembly, the Dana 44 was originally installed in the 1989 Turbo Trans Am 20th Anniversary Indy Pace Cars. They were also offered as over-the-counter assemblies and are the best factory rear ends offered for late-model Camaros. These units are unfortunately no longer available through GM, but some aftermarket companies still offer them. The only problem is the price tag—about $1,800.

Modifications—For most high performance purposes, the stock rear ends can be modified with aftermarket parts and pieces to handle more horsepower. Auburn Gear (Auburn, IN) manufactures several Posi-units that fit into the 7-1/2-inch 10-bolt rears and a variety of gears from 3.08 to 4.11 ratios are available from companies like US Gear/Strange Engineering (Evanston, IL) and Richmond Gear (Liberty, SC). As of this writing, there are not many gear options available for the Borg-Warner 9-bolt rears.

Camaros. One of the first was the 10-bolt, Saginaw rear which uses a 7.50-inch ring gear. This rear was updated in 1985 to handle more horsepower by using a slightly larger 7.625 ring gear and 26-spline axle shafts. (Both of these rears are referred to as '7-1/2-inch 10-bolts' as a means of differentiating them from the earlier, and stronger 8-1/2-inch 10-bolt assemblies.) 1990 10-bolt rear ends were then upgraded to 28-spline axles.

Borg-Warner—Some high performance Camaros built during the 1985-92 model years were equipped with a Borg-Warner rear that is easily identified by a nine-bolt rear cover. These beefier units use a 7.75" ring gear and bearing retainer plates instead of the standard C-clips.

One of the easiest ways to improve acceleration is to install a lower ratio (higher numerically) ring-and-pinion set. A fairly good selection of ratios is available through companies like U.S. Gear/Strange Engineering. Photo by Dave Emanuel.

Aftermarket Choices

If you are planning on making some serious horsepower with your Camaro, check into aftermarket rear ends that are bulletproof and can definitely take the abuse of 450+ horsepower engines. The trusty 12-bolt Chevy rear has been the workhorse of many drag racers, and Summers Brothers (Ontario, CA) manufactures a heavy-duty 12-bolt for Camaros. This rear features 32-spline rear axles and a hefty torque arm. The Summers Brothers 12-bolt is made to fit high performance IROCs and Z/28s that use factory rear disc brakes, but with some modification, you can fit this unit with JFZ or Wilwood rear disc brake kits if desired.

Ford 9-Inch—Although installing any Ford product on a Chevy is grounds for lynching among Chevy purists, there is one exception that should be considered—the Ford 9" rear end. It's by far one of the strongest rear ends for racing and high performance use. Ford 9" rears, modified to roll beneath a 1982-92 Camaro, are available from manufacturers such as Strange Engineering, (Evanston, IL) Currie Enterprises (Anaheim, CA) and Moser Engineering (Portland, IN). Although these units are very strong, they're expensive because they are filled with race quality components (this is especially true of the Strange Engineering assemblies which are suitable for any type of racing).

The Ford rear axle offers a number of advantages—in addition to its tremendous strength, it has a removable center section and the widest assortment of replacement gear ratios available. Take a peek beneath any NASCAR Winston Cup race car and you'll see a Ford 9" rear axle. For any serious race or Pro Street car, it's the best way to go. At least Fords are good for something!

Once you get the drivetrain worked out, you'll need to get the power to the ground. A good set of slicks will be tremendously helpful during quarter-mile assaults. Photo by Dave Emanuel.

The secret to high performance handling is no secret at all. The only requirement is to have a handling package in which all the components are matched to each other and work well together. This 1992 Project Camaro from All Chevy Magazine generates over 1.3 g's on the skid pad! Photo All Chevy Magazine.

SUSPENSION 9

Your late-model Camaro is already one of the best handling cars on the road today. The Camaro's chassis design has always defended itself admirably on twisty mountain roads. The front suspension utilizes a modified MacPherson strut, with a separate lower control arm and coil spring arrangement, and the strut replacing the upper control arm and conventional shock absorber. Unlike conventional MacPherson struts, the Camaro's coil springs do not wrap around the struts, but rather stand between a single lower A-arm and crossmember. In the rear, the Camaro utilizes a torque arm rear suspension with a Panhard rod and coil springs and shocks. With this arrangement, the driving torque is absorbed by the transmission mount. From 1982 to 1992, this basic chassis design has been used, with various adjustments made in shock valving, anti-roll bar size and coil spring rates between models and model years.

Race cars like this SCCA Trans-Am Camaro, driven by Jack Baldwin in the SCCA Trans-Am series to win the 1992 Championship, use sophisticated suspension systems to provide cornering forces of up to 2 gs.

There are a number of relatively inexpensive methods you can use to increase the handling performance of your Camaro, depending on your intended use and budget. A Camaro can be made to corner over 1.0 g, but you'll sacrifice quite a bit of ride comfort in doing so. Therefore, before you begin investing in an entirely new chassis setup, carefully evaluate what type of driving you're most likely to do. In most cases, changing the tire pressures and setting proper wheel alignment will bring noticeable improvements and cost little. Or, you can lower your Camaro, replace the front control arms, replace bushings and put in beefier anti-roll bars for superior autocross handling. It doesn't make much sense to spend thousands of dollars building a killer suspension for a car that's driven back and forth to work and only occasionally allowed to strut its stuff. So make improvements in stages, starting with the easy, inexpensive modifications, until you reach the level of handling you are most comfortable with.

Most of the recommendations that follow are done without getting into too much detail on suspension geometry. The principles involved in how and why a car corners are exceptionally complex. If you want to dive into the mystery of suspension geometry, then I recommend you read HPBooks *Chassis Engineering*, by Herb Adams; HPBooks *How to Make Your Car Handle*, by Fred Puhn; and *Performance Handling* by Don Alexander, available from Classic Motorbooks. Now let's get on with some basic improvements.

WHEEL ALIGNMENT

One of the easiest and most inexpensive ways to improve handling is by changing the factory alignment settings of the Camaro. The factory has set a specific wheel alignment that is a compromise between extended tire wear, a comfortable ride and handling performance. The front wheels of your Camaro have various alignment adjustments—caster, camber and toe. Each of these can be adjusted to improve handling. Although some of these adjustments can be made yourself, it is best that you take them to an alignment shop (just about any large tire store will have a professional setup) and ask them to reset your Camaro's alignment to the following specs.

Caster

Caster is the inclination of the steering axis and its setting affects a car's tracking characteristics. Increasing the amount of caster (positive) tends to keep the wheels pointed straight ahead and increases the amount of steering effort required to turn the wheels (as when cornering or changing lanes) from their straight-ahead position.

Camber

Camber is the most common adjustment made to increase handling performance. Viewed from the front, a car with *negative camber* has its wheels angled in (towards the chassis) at the top. If a vehicle has it wheels angled out at the top, it has *positive camber*.

For most performance applications, a bit of negative camber is more effective; a setting of 1/2 degree negative is recommended for Camaros with stock wheels and tires. Any more negative camber in a street car will cause excessive tire wear and is not recommended. Negative camber helps maximize tire contact patch because it compensates for the body roll and lateral weight transfer that results during cornering.

For example, suppose your Camaro has zero camber (which is about what it has stock) and you were driving down a curvy road at a fairly fast rate of speed. When you make a tight turn, the weight of the car shifts towards the outside of the turn, the body rolls in that direction and the tire is angled out at the top. At this point, the outside front wheel has been pushed into a positive camber position whereupon the inside portion of the tire's tread is lifted, so it's no longer contacting the road. A bit of negative camber is used to help offset this situation, and in racing applications where tire wear is less of a factor than cornering capability, significantly greater amounts of negative camber are dialed in.

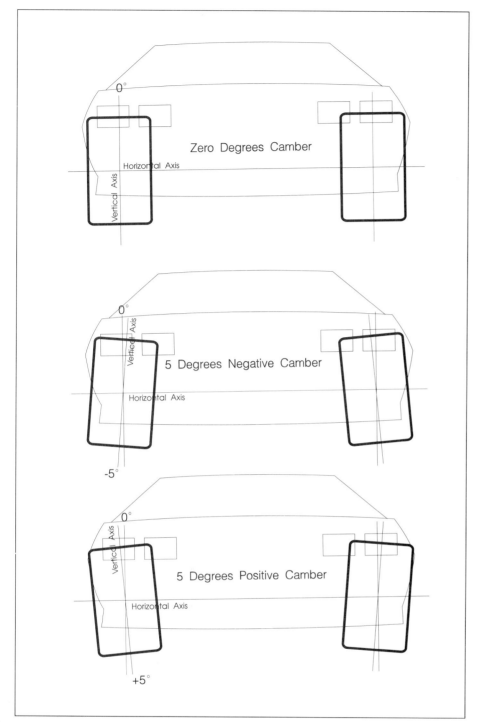

Camber is the most common adjustment made in the name of high performance. These three drawings illustrate differences between zero, negative and positive camber. for Camaros, a setting of 1/2 degree of negative camber is recommended for Camaros with stock wheels and tires.

For example, if you look at most shopping carts, you'll notice that the wheels are set back behind the pivot points of the cart. When you push the cart forward, the wheels shift rearward. This makes the cart easier to steer. Similarly, the caster settings affect a vehicle's steering. Compared to those for previous models, the caster specifications for third-generation Camaros may seem a little bizarre because a lot of positive caster is designed into the front suspension. To optimize cornering power, caster should be set between 4-1/4 and 5-1/4 degrees positive.

Toe

Toe adjustment is the angle at which the front of the tires are pointing. *Toe-in* refers to a setting which positions the tires so that they're angled inward in front—towards

109

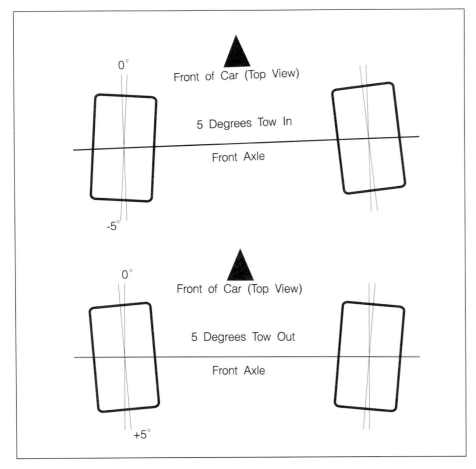

Toe is the angle at which the front of the tires are pointing. For maximum performance, you should set your Camaro with zero toe—in other words, with the tires pointing straight ahead.

each other. *Toe-out* is the opposite, where the tires are further apart in the front than they are in the rear (see illustration). For performance applications, a toe setting of zero is recommended.

TIRES

Tires have more effect on a car's handling than any other component, and they determine the amount of cornering power available. Tire technology has progressed significantly in recent years, and rubber compounds and tread designs have made tires much more stable at higher speeds and less prone to blow-outs. There is now a complete selection of tires ranging from bias-ply to radial. And the latest in directional and asymmetrical tires hug the road and displace water for traction far superior to tires of a decade ago. In fact, some tires are so sophisticated, that they are made for right or left sides only. Lower profile designs have also resulted in tires that respond to steering input much quicker for performance driving.

Increasing Tire Size

The 1982-92 Camaros may be wearing 14", 15" or 16" tires—it all depends on model year and options. Most of these, particularly those on the later model Z28s and IROCs, are excellent low-profile designs that produce good traction and cornering power. But further improvements can be made by simply increasing the tire width and size. For example, going from a 15-inch to a 17-inch series tire will yield a tremendous improvement in responsiveness. This is a result of a shorter sidewall which makes the tire stiffer and more responsive to steering wheel input.

The Plus System—Moving up to larger wheels and tires is a simple task that involves a little thought before you make a purchase. The thing to be aware of here, is not to go overboard with tires that may not fit in your Camaro. The easiest and safest way to move up to larger wheels and tires is to use a lower profile tire so that the tire overall diameter is no different from the stock tire diameter. This system is referred to as Plus 1, Plus 2, and Plus 3, a concept that enables wheel diameter and tread width to be increased 1, 2 or 3 inches while tire sidewall height is reduced, without altering ground clearances or speedometer accuracy. Increasing the diameter of the wheel while decreasing the sidewall height of the tire results in improved lateral stability, quicker steering response and shorter braking distances. It is best that you decide which brand of wheel and tires you want first, then consult with the manufacturer. Most have technical hot lines to answer any questions.

Selection

The type of tire you select is less important than selecting the proper size, namely because there are so many excellent types available for third-generation Camaros. Manufacturers like Goodyear, Bridgestone, Pirelli, BFGoodrich and Yokohama are popular brands with tire models and sizes that complement Camaros. Because these companies develop new tires and compounds continuously, there's little point in recommending a specific

One of the best ways to improve handling and straight-line traction is to use wider and stickier tires. Current tire technology has yielded street tires that offer as much traction as the slicks of a few years ago. A unidirectional tread is especially useful for maximizing cornering power.

tire make. But remember, handling is partly a subjective evaluation, so "feel" is an important consideration. A tire that provides the best road feel may not offer the best handling characteristics.

It should also be noted that once you're in the land of high performance tires, small improvements in handling come at very large expense. Unless you plan to do a lot of competitive autocrossing or make late night banzai runs, you don't need the absolute best tire available. By moving down the ladder a rung or two, you may find tires that are more than adequate for your needs but cost less.

WHEELS

Wheels are commonly replaced for two reasons—to accommodate a change in tire size and for cosmetic reasons. Not only can you add a design that significantly alters the appearance of your Camaro, but you can also increase performance. But when changing wheel size, fit and clearance is a primary consideration. Determining the best wheel size is most commonly done by measuring the backspacing of your wheels.

Backspacing

Backspacing is the amount of wheel offset needed to properly center the wheel in the wheelwell. For example, if the wheel measures 8-inches in width, and the backspacing is 4-inches, then the mounting surface of the wheel is exactly in the center. Some wheels need a different backspacing to fit correctly. With the right backspacing, you can easily swap in a set of wider and larger tires for improved handling.

Wheel Selection

When selecting wheels, you should first be sure of the size and backspacing you are going to need. For optimum handling, make sure you purchase a good, quality wheel that will last a long time and be able to withstand a lot of abuse. Most of today's performance wheels are made from aluminum. There are however, many ways they are manufactured. *Cast* wheels are cast in one piece. *Modular* wheels typically utilize front and rear rim halves that are welded or riveted together. The addition of a cast center section forms a three-piece modular wheel. *Composite* wheels use a rolled one-piece rim with a welded cast center. *Billet* wheels use a rolled one-piece rim with a welded-in center machined out of solid billet aluminum.

A good high performance wheel is light, and strong enough to take the forces encountered with high speed handling. Specifically for Camaros, the best bet for a good wheel is a billet or a modular three-piece wheel. Depending on the manufacturer, these wheels can run anywhere from $250 a piece to $800 or more for each wheel.

For most applications, Camaros can accept up to a 9-1/2" wide wheel for a 10" wide tire that will fit in the stock wheelwells, provided backspacing is 4-1/2" to 4-3/4". But before laying your money down, be sure to measure your Camaro or have a wheel dealer do it for you. Rear dimensions may vary if you have drum or disc brakes. Additionally, disc brakes in the rear may cause problems with the tire rubbing against the parking brake cables. Camaros with drum brakes

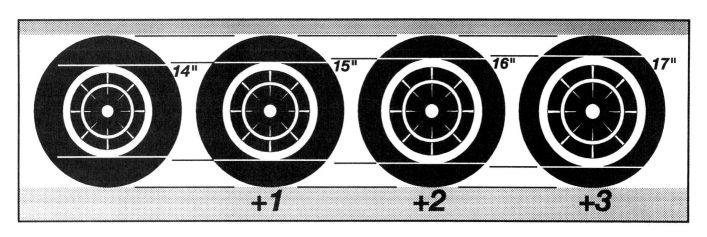

In addition to improving appearance, custom wheels can also improve handling. This Jongbloed wheel is 17" in diameter and 9" wide and provides a home for a Goodyear GSC 255/45/ZR17. This is a standard Corvette tire and on a Camaro, it greatly increases cornering performance.

on the rear should be able to handle 10" wide wheels with 4-3/4" backspacing. Some manufacturers use metric backspacing measurements and recommend using a 9-1/2" or 10" wide rear wheel with a 10 or 11 mm offset.

COIL SPRINGS

The purpose of suspension springs is to hold the car steady while allowing the wheels to follow road irregularities. In general, the softest possible springs will do this job best. Softer springs will allow each individual wheel to move in relation to the chassis while having the minimum effect on the driver's compartment. This translates into a soft ride, noise isolation and good handling.

Many enthusiasts believe that swapping in a set of stiffer springs will make the car handle much better, but in reality, all they really do is make the ride harsher, and they do not make much of an improvement on handling. As long as the springs on a car are stiff enough to keep the car from bottoming out, they are adequate. If a car is lowered, a slight increase in spring rate can be used to compensate for the reduced ride travel.

Some car enthusiasts have the mistaken belief that if 300 lbs-in. coil springs are good, then 600 lbs-in. springs have to be better. They're wrong. Optimum road-holding demands that the tires be in contact with the pavement; a soft spring lets the wheels follow road irregularities for maximum adhesion.

Springs greatly affect the ride, handling and ride height of a vehicle. Although most springs for a specific vehicle look nearly identical, there can be significant differences in wire diameter and overall height. High performance springs, like these from Eibach, can provide superior cornering while simultaneously lowering ride height.

CUTTING COIL SPRINGS

Before you invest in a new set of coil springs to lower the front end of your Camaro, you may want to consider cutting them instead. According to chassis expert Herb Adams, cutting your Camaro's coil springs by 1/2 a coil will lower your car but still maintain adequate ride height.

1. Carefully remove the coil springs following the procedure outlined in your car's shop manual. Note how the end coils of this stock front spring are bent slightly so that it seats properly in the frame and A-arm. Most stock springs can be trimmed a half coil to provide the proper ride height for improved handling. Mark the spring directly across from the original end as shown.

2. Cut the spring with an acetylene torch.

3. Heat the half coil below your cut so you can bend it to match the spring's original shape.

4. Quickly turn the spring upside down and bend the top coil by pushing down on the spring. DO NOT quench the spring with water; allow it to air-cool slowly.

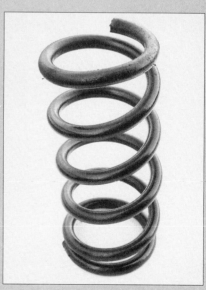

5. Paint the spring and reinstall it according to the directions in your shop manual. Realign the front suspension following the recommendations in VSE's Performance Handbook.

Center Line billet aluminum wheels add a distinctive drag race appearance when mounted on a Camaro. This is especially true if relatively narrow front wheels are combined with wide rears. This Camaro has 6"-wide wheels on the front and 10"-wide wheels on the rear. Surrounding the wheels are Mickey Thompson Sportsman tires on the front and ET drag slicks on the rear. Photo by Dave Emanuel.

Some suspension experts, such as Herb Adams, recommend factory springs for general use. For street use, you can trim one half coil off the top of the spring with an acetylene torch to lower the car slightly.

Most front spring rates range between 300 and 350 lbs-in. But the Camaros have rates of 365 lbs-in. for base coupes, with IROC-Zs and Z28s with rates of 548 lbs-in.! Testing by Herb Adams has shown that the ride can be greatly improved by installing softer springs, with no effect on the handling capabilities of these cars. Many enthusiasts buy the standard rate soft springs from their dealer for

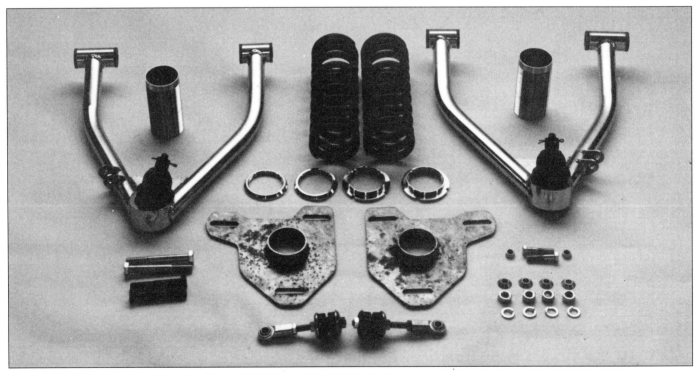

Art Morrison offers this front-end suspension kit that includes coil springs that fit over the stock front struts. This arrangement allows ride height to be easily adjusted. Photo by Dave Emanuel.

One of the best ways to lower a Camaro is through use of a custom lower control arm. This one from Chisolm Enterprises can lower a stock Camaro as much as 2 inches without affecting suspension geometry and ride quality. This control arm is also a few pounds lighter than a stock one, which reduces the unsprung weight.

Adjusting Ride Height

Although the spring rate will not change during the life of a spring, the spring load can change. This is commonly called *spring sag*. Loss of load, or spring sag, can be caused by a variety of reasons, including poor metallurgy, overloading and even fatigue due to high mileage. If a spring has lost load, this problem should not be corrected by changing the spring rate. If a spring has sagged, you can get the spring load back to normal using several methods. For coil springs, the usual procedure is to place a rubber shim on top of the spring to increase the load (shims are available through car dealership parts departments and from auto parts stores).

Lowering—The least expensive method of lowering is to simply cut your existing springs.

Your car's existing springs have already taken a permanent set, so you know where they will end up. If you want to lower your car, you can

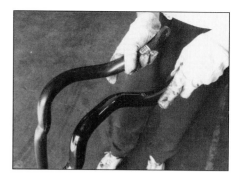

Anti-roll bars (also called sway bars, stabilizer and anti-sway bars) improve handling by increasing roll stiffness. The bars from an IROC-Z or Z28 are the best, even for some racing applications. The stock bar is hollow while most aftermarket manufacturers use a solid bar (right) which is stiffer but considerably heavier.

cut your existing springs and achieve the ride height you want with one operation.

Springs don't wear out, so you can save the cost of new springs if you trim your existing ones. If you want to feel like you bought new springs, paint your old ones. We have found that most coil springs should be cut 1/2 coil to lower the car and still keep adequate ride height.

Aftermarket Springs—Some manufacturers such as Eibach, Chisolm, Suspension Techniques and Air Camaro offer springs that provide a 1-1/2" to 2" drop in ride height. These springs make for a slightly rougher ride but dramatically improve handling and appearance. High performance springs come in two types—*progressive* and *linear*.

Progressive springs become increasingly stiffer as the load applied increases, but not in direct proportion to that load; a doubling of the load applied to a progressive spring does not double its stiffness (spring rate).

Linear springs provide more consistent stiffness when compression forces are applied. They compress at a steady rate as cornering loads increase. Stiffer springs are normally used for racing applications, where cornering forces are high and ride quality isn't important. For high

these cars and cut 1/2 coil for the proper ride height. Trimming the coils by 1/2 a coil will increase the rate approximately 10%. But the true purpose of trimming the front springs is to lower the car for improved aerodynamics and better handling, not to increase the spring rate. It also gives it a nice look.

If you intend to change front struts, beg, borrow or steal an impact wrench. It's virtually impossible to loosen the top nut without one. Photo by Dave Emanuel.

Weight transfer plays an important role in cornering. Note how this Camaro almost has the left front tire off the ground as it makes a hard left. Here, the weight is transferred to the outer right wheels by centrifugal force--with the help of an anti-roll bar.

performance street applications, many spring manufacturers recommend linear springs.

Dropped Spindles—Another popular method of lowering is using a modified spindle. This works great for most Camaros but you should verify that any non-stock spindle is compatible with the front brakes. Also note that if the front is lowered, some attention must be given to the rear to keep the car at or near level.

ANTI-ROLL BARS

As mentioned earlier, as a car goes around a turn, the body will roll and effectively change the camber angle of the tires in relation to the pavement, to a positive camber condition. To dial in enough negative camber to compensate would mean that your tires would wear quicker for daily use. The best way to increase the *roll stiffness*, therefore, is to increase the size and effectiveness of the anti-roll bars. Anti-roll bars limit the roll angle of the car by using their torsional stiffness to resist the movement of one wheel up and one wheel down.

Increasing Size—Late-model Camaros are already factory-equipped with beefy anti-roll bars. The late-model IROC-Z, RS and Z28 models have a 1-3/8" front bar and 7/8" rear bar. For most street use, these stock bars are very adequate. But if you are thinking of doing some serious cornering, you may want to purchase a set of aftermarket anti-roll bars.

The stiffness of an anti-roll bar increases very quickly as its diameter is increased. The stiffness is a function of the diameter to the 4th power, or D^4. This means that a 1-1/4-in. bar is 2.44 times as stiff as a 1.00-in. bar. Also to be considered is the length of the anti-roll bar arms. The longer the arms, the less effective the bar will be. For example, an anti-roll bar with 6.00 inch arms will produce twice as much roll stiffness as a bar with 12.00 inch arms (provided they are of equal diameter).

Be aware that increasing rear anti-roll bar stiffness while leaving the front bar unaltered can lead to an oversteering condition. If excessive, this can prove very dangerous as cars setup to oversteer often have a nasty habit of swapping ends when

One of the easiest ways to improve the rear suspension is to substitute polyurethane bushings for the stock rubber ones. These rear lower control arm bushings are from Energy Suspension. The company also offers Panhard rod and front control arm bushings. Photo by Dave Emanuel.

driven through a turn at high speeds. Like most other automotive systems, chassis tuning involves many factors that must be balanced for optimum results.

SHOCK ABSORBERS

The purpose of shock absorbers is to control the velocity of the suspension. If the shocks don't have enough resistance, the spring will move the suspension too fast and it will have an under-dampened motion. If the shocks are too firm, the motion will be over-dampened. It is important to have just the right amount of dampening to control the spring action of the suspension in order to maximize tire contact with the road.

Proper Dampening—Extra firm shocks have the same negative effects on ride and handling as extra stiff springs. The tires cannot follow the road irregularities unless they are free to move in relation to the chassis. The relative motion must be as free as possible, but it also must be controlled. The problem is how to get a stiff enough shock to improve handling, yet soft enough to keep the ride smooth. The answer is an adjustable shock.

Adjustable Shocks—Adjustable shocks allow each car to be tuned for critical dampening, which is especially important if you change the spring rate of your coil springs. Changing spring rates can have a significant effect on the stock shock absorber's dampening ability. Because a performance spring is stiffer, a stock shock will not have enough stiffness or strength to control the spring oscillations. Run your shocks as soft as possible—just enough so the car doesn't wallow over bumps.

Make sure you purchase adjustable shocks that will work for a wide variety of situations. Some are available with adjustable rates that offer better performance for street driving, yet can be easily adjusted for autocrossing. Manufacturers such as Bilstein, Tokico and Koni make performance shocks with this range. All of these manufacturers have extensive research and development departments, and they have all participated in motorsports with great success. Your local performance parts dealer should have literature on each brand.

The Camaro's rear suspension is easily modified because most of the components are well within wrench range. For serious running, the Panhard rod and lower control arms should be reinforced or replaced with performance units. Photo by Dave Emanuel.

BUSHINGS

Suspension bushings are deceptively simple devices, because although they seem insignificant, they do play a very important role in your car's handling.

Under most driving conditions, rubber bushings are the best choice. However, under high performance

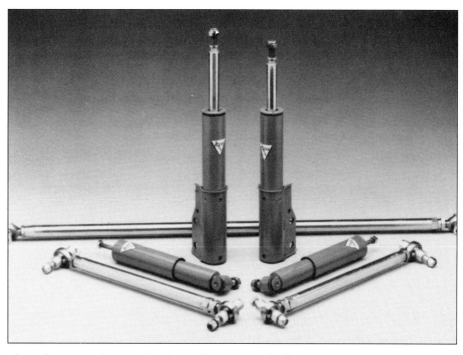

The stock rear control arms and Panhard rod are notorious for flexing under hard acceleration and braking. Several companies offer tubular replacements with Heim joints. Although designed for racing, these components can be used for street-driven vehicles. This particular kit is packaged with Koni front struts.

Improved stopping power should be part of any high performance modification program. Several brake options including replacement calipers, are available.

When converting from standard to 1LE brakes, the stock spindles must be swapped for 1LE spindles--which move the wheels outward 1/2-inch.

Upgrading a Camaro brake system to the Corvette-type normally found on 1LE Camaros is easy. The hard part is finding the pieces in a wrecking yard.

driving conditions, the bushings will bend or *deflect*, which results in a change in camber to positive, which reduces cornering power of the tire.

Replacement Bushings

There are several different kinds of replacement bushings available. Which you choose depends on your tolerance for noise, ride harshness, and your intended use.

Urethane Bushings—In recent years, several companies have offered urethane suspension bushings. They are effective in reducing deflection, and improve handling—many enthusiasts swear by them. By comparison, rubber bushings allow rotation by the internal shear of the rubber itself. This means there is no sliding motion between any of the members. The rubber flexes to allow the inner and outer sleeves to rotate relative to each another. This *deflection* causes improper suspension geometry, so a firmer bushing is required. Since there is no sliding motion, there is no friction-caused wear and no need for lubrication. Because the rubber is molded to the inner and outer sleeves, there are no critical tolerances to maintain during manufacturing. This is one of the features that allows rubber bushings to be made inexpensively.

When a urethane suspension bushing is used, the bushing material cannot deflect. Many people who buy aftermarket urethane bushings believe that the urethane causes squeaks. Some manufacturers have eliminated squeaks by several methods. Greasing urethane bushings before assembly works until the grease is forced out or washed away. Once the grease is gone, the urethane is again able to bind to the steel sleeves, and the driver hears squeaks and moans. Without lubrication, the urethane can stick to the steel, and the suspension does not move smoothly.

Some urethane bushings are customized to provide better service by hand-fitting the bushings to obtain proper clearances and installing grease fittings for regular lubrication, which is what many manufacturers now do.

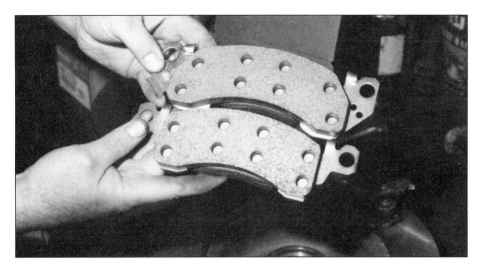

Several grades of brake pad material are available and each offers different stopping characteristics. Premium quality semi-metallic pads offer excellent long-term resistance to fade and longer life than "cheapie" pads.

Nylon Bushings—Nylon bushings, when properly supported and lubricated, have proven themselves in many industrial applications. Nylon inserts are inexpensive and are available from any bearing supply house. These "nyliner" inserts require sleeves machined to close tolerances for proper operation, so the cost of installing nylon suspension bushings is slightly higher than urethane bushings. Compared to the cost of labor to remove and replace the control arms, however, the amount of increase is insignificant.

Properly made sleeves to support nyliner bearings should have zerk fittings for easy lubrication. Cars have been known to run more than 100,000 miles on the same set of nyliner suspension bushings. All bushings should have zerk fittings so they can be greased every six months.

Ride & Noise Changes—If the suspension bushings you install are free to rotate without binding, the ride characteristics of your car will not change much. However, both cornering power and steering response will improve substantially.

PANHARD ROD

Late-model Camaros utilize a device called a Panhard rod—a bar located just above the axle, running from the axle housing on one side to the body on the other. What this suspension component does, is to keep the axle from shifting from one side to another under hard cornering conditions. This device can also be improved upon with an aftermarket piece. The stock Panhard rod is made up of a U-shaped piece of

If you're looking for maximum braking power, an aftermarket brake kit is the answer. Brakes such as this are used on a variety of road racing cars, but they can be used on a street car.

The factory lower control arms are great for stock performing Camaros, but for a high horsepower car, you will need something beefier.

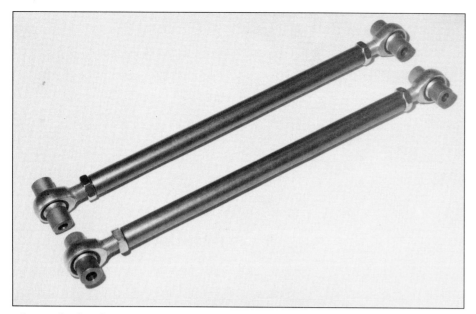

These Herb Adams lower control arms are solid mount type that are lightweight but further decrease your Camaro's ride.

steel with large rubber bushings at each end. Under hard cornering forces, such as those applied with larger wheels, tires and stiffer springs and shocks, the stock rod will sometimes bend and deflect, causing some rear end wobbling and possibly some tire rubbing on the fender well.

Manufacturers such as Dick Guldstrand and Herb Adams make a replacement Panhard rod which is made of a solid piece of steel with adjustable Heim joints at each end. This will keep the axle located at the center of the body during hard cornering when used with a performance suspension.

BRAKES

Improving on your Camaro's braking system is a natural thing to do, if you plan to make suspension and engine modifications. If a car can accelerate like a rocket and corner like it's on rails, it must also be able to stop on a dime and deliver eight cents change

Stock third-generation Camaro disc brakes are of single piston caliper design and are adequate for most situations. The 1LE owners are luckier in that they have the larger, Corvette-type, aluminum calipers with the larger diameter rotors. Some aftermarket manufacturers offer performance brake pads made of semi-metallic material, which improves braking performance. But to really improve your Camaro's braking, you can order a set of the aluminum caliper brakes and 12" diameter rotors used on the 1LE from your local Chevy dealer, although they will be expensive.

A less expensive alternative is a set of aftermarket brakes available from JFZ or Wilwood. These brake manufacturers make a system for late-model Camaros that feature a dual-piston caliper and rotors up to 13 inches in diameter. This combination offers a vast improvement of the braking ability of the stock system. These brakes are supplied in kit form and are very easy to install.

Rear Discs

If your Camaro isn't already equipped with factory rear disc brakes, installing an aftermarket set will greatly decrease your Camaro's stopping distance. The only thing you have to remember however, is

The Herb Adams bars mount solidly to the frame and axle, providing increased traction and virtually eliminating wheel hop.

Here's a shot of All Chevy Magazine's project '91 Z28 Camaro. It features the solid Herb Adams lower control arms, subframe connectors and a Herb Adams Panhard rod.

to change over to the proper master cylinder that operates front and rear discs. If you add an aftermarket rear disc brake kit without changing the master cylinder to match, your brakes will feel sluggish and you will have to press very hard on the pedal to stop.

Most of the aftermarket brake manufacturers will supply you with an additional proportioning valve which will also take care of a mismatched master cylinder. These are available from JFZ, Wilwood and GM. With all the right pieces and an improved braking system, you will definitely enjoy the way your Camaro stops.

SUBFRAME CONNECTORS

One of the flaws of the unibody style chassis is the elimination of an underbody frame. Unfortunately, the Camaro utilizes this type of construction. Over time, the constant flex and bending of the body will cause problems such as improper door alignment, squeaks and rattles and T-tops that all of a sudden don't seem to fit properly.

There are several manufacturers who make subframe connectors that connect the front half of the unibody chassis to the rear half, which greatly strengthens the Camaro's chassis. Subframe connectors help improve the car's handling as well as keep the chassis stiff enough so that other parts of the body will not flex or bend under acceleration, cornering or rough road conditions. Depending on the manufacturer, the subframe connectors can be bolted or welded on. Air Camaro is one manufacturer that sells a bolt-on subframe connector which can be installed by an average enthusiast in one afternoon. South Side Machine is another manufacturer that sells a weld-on subframe connector. Either type of connector works well and both are relatively inexpensive. Installing a subframe connector is an essential part of a performance Camaro's handling and will also lead to its longevity.

LOWER CONTROL ARMS

When it comes to straight line acceleration, there's no substitute for traction. Although you might think that tires play the most important factor in obtaining traction, it is also the chassis' role to help keep the tires planted on the pavement.

Late-model Camaros use a simple rear suspension system that can increase traction. It is comprised of a solid rear axle that is supported by coil springs, shock absorbers and two lower control arms. These lower control arms keep the rear axle from twisting during acceleration and increase the traction of a stock Camaro. If your Camaro is modified however, the stock control arms might not do the job.

The weak point of the stock control arms is the use of large rubber bushings. The control arms are connected from the bottom of the axle housing to the rear subframe, in front of the rear wheel. With large rubber bushings and a performance engine, these factory control arms can flex, resulting in excessive wheel hop during acceleration. There are several options you can take to eliminate this problem.

The first is to remove the factory control arm bushings and replace the rubber bushings with urethane. The factory arms are made from U-shaped steel and they can be strengthened further by welding in a piece of steel at the bottom of the arm to make it a solid, box-type tube.

Aftermarket Choices–The next step would be to purchase aftermarket lower control arms. Manufacturers such as Hotchkiss Performance make a solid steel lower control arm that uses urethane bushings lubricated by zerk fittings installed on the bar. These provide a good, solid link between the axle and frame that virtually eliminates wheel hop. Herb Adams also manufactures lower control arms but these use Heim joints at each end of the bar. The advantage to these solid bars is that they are lighter in weight, stronger and capable of handling very high horsepower requirements. The drawback however, is that since they are solidly mounted to the axle and frame, road and noise vibrations are transferred into the interior of the car. The ride is also a bit rougher but for a high performance Camaro that runs 12-second quarter-mile times, the slight discomfort is well worth the increase in traction.

On All Chevy Magazine's project Camaro, the installation of the Herb Adams lower control arms lowered the quarter mile ETs by a half a second!

INSTALLING SUBFRAME CONNECTORS

Installing a set of subframe connectors on your Camaro, will increase its handling, acceleration and eliminate any excessive body flex. The difference is dramatic. Subframe Connectors will also eliminate any door and T-top problems in the future.

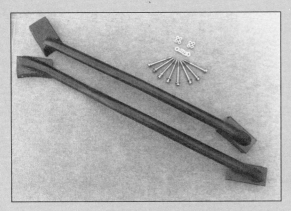

1. Here is a typical set of subframe connectors that are of the bolt-on variety. These are simple to install and are made of high grade steel tubing for added strength. These particular connectors are available from Pro Chassis (Hacienda Heights, CA).

2. The subframe connectors are mocked on, and a punch is used to indicate where the bolt holes need to be drilled.

3. In some later model Camaros with dual catalytic converters, it may be necessary to temporarily unbolt the cat-hangar to have enough room to bolt in the subframe connector.

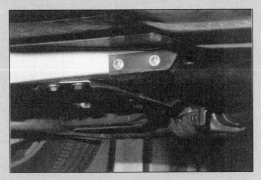

4. Once the bolt holes are drilled, the subframe connectors can be attached to the front and rear of the Camaro's unibody chassis.

5. This vital link between the front and rear will dramatically increase your Camaro's cornering, acceleration and eliminate any flexing.

While most people equate performance with horsepower and handling, some prefer to measure the term in decibels. This specially prepared Camaro is used to demonstrate the performance of Alpine stereo systems and components. Stereos, seats, roll bars, and steering wheels are some of the interior modifications you can make to personalize your Camaro.

INTERIOR MODIFICATIONS 10

If your Camaro is going to be a high performance street machine, it had better feel like one when you're in the driver's seat. It is important to feel comfortable and in control during high performance driving, and that comfort begins with the seat and steering wheel. Of course, it also helps to have excellent sound to cruise by, and because Camaro's are popular among car thieves, you'll want added protection for your investment. To have the ultimate Camaro, a few basic interior modifications are in order.

If the need for speed has prompted you to undertake engine and chassis modifications, installation of a realistic speedometer is advisable. This 140 mph speedometer from Classic Camaro, fits '82-'87 models and is easy to install.

A custom steering wheel with a nice fat grip eases hand fatigue on long trips and allows better road feel and control. This sport wheel from LeCarra features a four-spoke design with a black leather wrap.

are designed to fit the contours of your hands while others have the grip placed for the proper hand positioning while driving.

Whichever type of wheel you choose, make sure it comes with an installation kit designed for a Camaro. The actual installation is very simple and most kits will work with tilt or standard steering columns.

STEERING WHEELS

A top-quality steering wheel is essential for control under high performance driving conditions. A good replacement steering wheel can cost anywhere from $100 to $800 depending on the style and the manufacturer. However, if you have a 1990-or-later Camaro, it is equipped with a driver's side supplemental restraint system, otherwise known as an airbag. If that is what you have, then you should definitely keep the wheel. Trying to remove the restraint system is not in your best interests. Fortunately, 1990-and-up Camaros come with excellent steering wheels, so there's little need for a new one.

For good steering feel and control, any good, thick, leather-wrapped wheel is suitable; there are many styles and sizes available. Late-model Camaros come from the factory with a 14" steering wheel, and most aftermarket wheels are about the same size. Some steering wheels

SEATS

One of the biggest shortcomings of the third-generation Camaros has been the stock seats. With such excellent factory stock handling capability, the Camaro's seats are seriously lacking in the areas of lateral and lumbar support. Increase the g factor and you'll be sliding all over the place. If you are serious about high performance driving, whether for autocrossing or some other form of motorsport, consider replacing at least the driver's seat with a performance aftermarket unit.

Top names in the custom seat business include Corbeau, Recaro,

Steering wheels with three spokes are also popular and fit well in a Camaro interior. However, custom steering wheels can't be installed in vehicles equipped with a driver's side air bag.

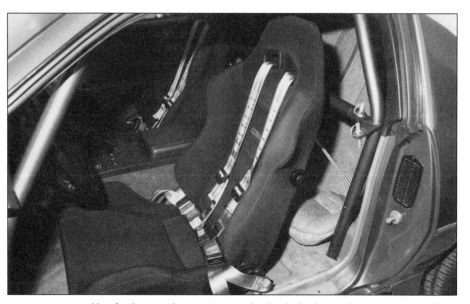

A custom interior adds a finishing touch to any Camaro that has had substantial improvements made to its engine, chassis or suspension. The driver and passengers in this 1991 Z28 have been treated to performance seating from Cobra, along with a full roll cage and Schroth safety harnesses. Photo courtesy All Chevy Magazine.

INSTALLING A GRANT GT PERFORMANCE STEERING WHEEL

Photos By Carl Caiati

1. The first step in a custom steering wheel installation is to pop off the center cover and remove the center nut. A wheel puller should be used to remove the steering wheel from the column.

2. With the wheel removed, the base plate remains on the column with the horn connection tube.

3. Wheel installation kits should provide a new horn wire which attaches to the factory connector.

4. Aftermarket wheels need to use some kind of adaptor or hub like this one which slips into place over the steering shaft.

5. This kit uses a dust cover that fits over the hub and covers the space between the column and steering wheel.

6. The aftermarket wheel can then be connected and bolted in place with the factory bolts. Make sure to pull the horn wire through all of the steps.

7. The steering wheel nut can then be reinserted and should be tightened snugly to eliminate any play in the wheel. Some kits will supply recommended torque settings.

8. The horn spring is then attached to the center shaft and the horn wire is plugged into the horn button which snaps in place over the spring.

9. The finished installation looks great and is a significant improvement over the stock wheel.

Many Grant steering wheels may be installed in conjunction with a billet-style adapter. Available in polished natural finish or with black anodizing, a billet adapter hub adds a high-tech finishing touch.

Flowfit and Cobra. The most important thing to look for is the width of the bottom portion of the seat. In many cases, seats with side bolsters that cradle your hips may interfere with the center console. Before you purchase any seat, take measurements of your stock seat's width and make sure the aftermarket seat will come close to those dimensions. You should also be sure an aftermarket seat comes with proper frames or adapters that will mate to the existing bolt hole locations. Many seats don't come with any frame, in which case one will have to be fabricated. Only four bolts hold the stock seat to the floor, so fabricating a frame isn't a particularly difficult job. Just make sure you use SAE grade materials for the highest quality and strength. You wouldn't want to have the seat come loose in case of a collision.

For the most part, seat upholstery and adjustability determine price. More expensive seats are made of better materials and offer the widest variety of support and adjustments.

For everyday street driving, it's best to stay with a fold-down seat rather than going with a racing type seat that has a fixed back. Racing seats look great and may be very light, but the hassle of climbing in and out of them may drive you crazy. Several manufacturers offer high-back fold-down seats that provide just as much support as an all-out racing seat.

Seatbelts

If you're changing the seat, make sure it is compatible with the factory seatbelt system. Even if it is, you should upgrade the stock system with a performance aftermarket unit if you intend to do some serious driving. Most aftermarket belts will be labeled "DOT Approved" (Department of Transportation). If they aren't, don't buy them unless they are certified by a noted racing organization. For ease and convenience, you can't go wrong with the belts made by Schroth. These belts are similar in design to a three-inch racing harness, except they are much more comfortable to wear and attach and unbuckle as easily as the stock belts.

Competition Belts—If you want to go all out, install a five-point belt system from Deist or Simpson. Make

One of the most important features in a Camaro interior is safety. These harnesses from Schroth are small, simpler to use than larger 3-inch racing belts and feature an anti-submarining control to prevent the passenger from sliding under the belt in the case of an accident. Harnesses such as this are often required if a car is to be raced or autocrossed.

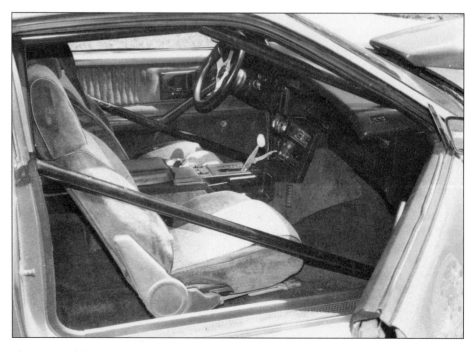

The interior of this Pro Street '84 Camaro features a performance steering wheel, aftermarket gauges and a full roll cage. Also note the performance-style shifter.

Window tinting is a popular addition in those parts of the country where sunshine and high temperatures abound. However, super dark tints, in addition to being illegal in some states, drastically reduce visibility at night. This can become a safety hazard. Installation of window tint film is best left to experienced hands.

sure that the harnesses are securely mounted to the floor (or other mounting areas established by the Sports Car Club of America or the National Hot Rod Association) using the proper hardware. Both the SCCA and NHRA are racing organizations that have rules and regulations regarding installation and mounting of safety equipment on many model sedans.

ROLL BARS

One of the best things you can do to increase the strength and rigidity of your late-model Camaro is to install a roll bar. The 1982-92 Camaros have some inherent body flex, which eventually loosens things and makes for a noisy ride punctuated by knocks and squeaks.

If you're planning to subject your Camaro to "severe service," a roll bar is a worthwhile addition—not only for safety but for its chassis-strengthening effect. Roll bars come in a variety of designs and sizes. The obvious choice is the one that is easiest to install and does not interfere with the rear seats or your driving space.

Unless you're going to do some competitive driving, a full roll cage that extends from the front to the rear of the car is not necessary. Most good, chassis-stiffening roll bars start from an area just behind the front seats and extend to a strong position in the frame of the rear of the car. The easiest roll bar to install is a simple four-point model that bolts right into designated frame points. Make sure that if you purchase a roll bar, that it meets the standards of the DOT, NHRA or SCCA.

If you're not sure where a roll bar should mount, you can obtain an SCCA rule book by writing or calling them at 9033 Easter Place, Englewood, CO 80112, 303/694-7222. The book shows the safest and most functional roll bar designs and locations for a roll bar for your Camaro, and clearly shows the proper mounting points.

If someone is building a custom roll bar for you, make sure they build it according to SCCA rules. If a roll bar is not installed properly, or in a proper location, it can cause severe injury to the driver or any other passengers in the event of an accident. An improperly designed roll bar can also be cumbersome when you or a passenger tries to get in and out. For this reason, it is always wise to select a good installer or a very simple bolt-on kit.

STEREO SYSTEMS

There seems to be an unwritten rule among some enthusiasts that a Camaro is not complete without a mega-watt stereo system that rattles the hubcaps off the nearest Mustang. The basic acoustical structure of the Camaro is excellent, as seen by the increasing quality of factory systems

An upgraded stereo system is always a worthwhile interior modification. Many aftermarket stereos offer a host of options that ensure top sound quality. This Alpine includes a CD player.

This Camaro has it all—a 400-watt Alpine stereo system, which includes 12 speakers, a compact disc player, cellular telephone and a unique Alpine alarm system.

throughout the third generation. But, although the factory system is better than most others, the cry for "more power" is often heard.

Stereo equipment can get expensive in a hurry once you start dealing in top-of-the-line components. But there is a budget approach that you may want to try first. There are several ways to add on to your existing system that you can try first before upgrading to a complete aftermarket system.

Receivers

To begin with, let's say you don't have a budget at all and you just want to improve your factory system. One of the first things you should realize, is that the factory Delco stereo system is a pretty good unit. As standard options in most 1982-and-later Camaros, the Delco stereo systems can range from a simple AM/FM without a cassette player to a sophisticated Delco/Bose series receiver that incorporates a high quality AM/FM/cassette with equalizer, tuned speakers and Dolby sound. Also available as an option on '89-and-later models is a Delco AM/FM receiver with a compact disc player. Unfortunately, since the Delco/Bose system also has higher powered, tuned speakers that produce exceptionally good sound, it was not available on convertible models because the soft top does not provide the proper acoustical environment for the system.

If you have the Delco/Bose system in your Camaro, you may want to keep it unless you are willing to spend $600-$800 or so on a better aftermarket unit. Except for some features, such as multi-station presets, built-in power and a removable chassis, the Delco/Bose receiver is better than most aftermarket stereo systems. Replacing it would actually decrease the level of sound and performance from the system unless a high cost, (over $600) high-end receiver were substituted.

If you're stuck with the base Delco receiver, and are on a limited budget, you may want to look at swap meets or local stereo shops for a Delco/Bose unit that was removed from another Camaro. The Delco/Bose unit is extremely simple to install and you'll have a much better unit at about half the cost of a good aftermarket stereo receiver.

Most Delco receivers are rather large, but they are self-contained and are easily reached by removing the front panel from the center section dash/console. When this panel is removed, it reveals the screws that hold the stereo in place. Once these are removed, the stereo can be gently pulled out and the wiring harness unplugged from the rear of the receiver. That's it. In most cases, the wiring harness for the stereo will plug right into the Delco/Bose unit without any problems. If not, you can purchase a late-model Camaro

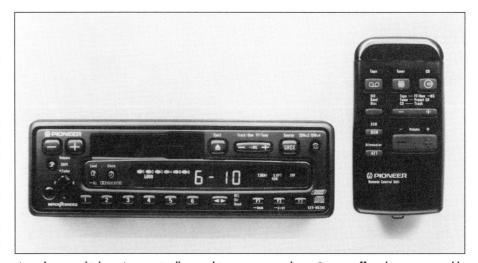

Just when you think you've seen it all, something new comes along. Pioneer offers this programmable receiver which features an AM/FM tuner, tape cassette, and a remote-mounted compact disc changer. Everything can be controlled by a hand-held remote control.

wiring harness kit for a couple of dollars.

If you really have your heart set on a high-end aftermarket receiver, make sure that it is able to fit in the stock Camaro stereo receiver location. In most cases, the width is not important since most receivers are the same, but the depth can be no more than 6-1/2".

Although many people think replacement of a car's stereo system can be done only by a professional audio installer, the stereo location on Camaros makes removal and replacement so simple that it's certainly a do-it-yourself project.

It's best to select a unit that will accept upgrades and add-on features. Unless you are purchasing an entire system all at once, look into a high-end receiver with the capacity to add amplifiers. This will enable you to use your existing speakers, and should you install better or additional speakers, you can always add amplifiers as required.

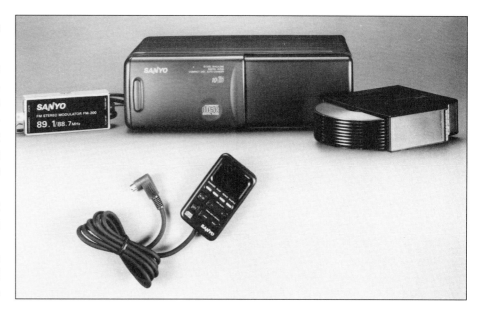

If all you want to do is add a compact disc player, this Sanyo unit may be the ticket. It can be mounted anywhere you please and controlled through the remote control key pad. Output from the CD is transmitted to the FM tuner on a preset frequency.

The best thing is to look at all the options available and decide which receiver will work best in your Camaro. You should also look at which features you are more likely to use. Many times you could end up paying more for a unit that has features you may not need.

Compact Disc Players & Changers

If you decide to add a CD player, choose one that is compatible with the receiver currently installed. Pioneer is one of many manufacturers that produces a great system that adapts to the stock GM receiver and works in conjunction with its own CD changer. Operating the Pioneer CD is simple with the remote control and the sound is transmitted through your factory stereo system. It also has the capability of programming your CDs and skipping from one CD to another.

Alpine also makes a CD adapter to work with stock Camaro receivers. The Alpine CD changer is also operated by remote and it has many unique CD functions, such as track search, program modes and random play.

Mounting–Mounting a disc changer such as the Alpine or Pioneer is very simple and there are

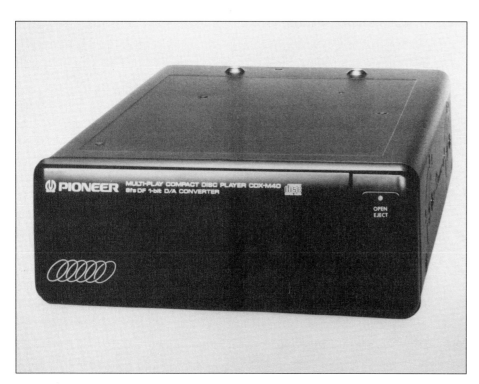

Compact disc systems can be integrated into the receiver or mounted separately. This Pioneer compact disc changer holds six discs and can be mounted in any convenient location.

several suitable locations. One of the most popular is in the back hatch area, behind the rear seat. Other suitable locations are under the front passenger or driver's seat. Many enthusiasts have noted that the Pioneer unit is better at absorbing shock from a bumpy road than most other units.

For general applications, choose an area that remains covered from exposure to the sun and is relatively vibration-free. Some competition vehicles involved in "sound-offs" mount CD changers in custom locations, such as directly behind the rear, fold-down seat. Others have been mounted in the rear where the spare tire used to be. Either way, it depends on your budget as to where you want to mount a CD changer.

Obviously, custom jobs require an installer and cost more. In most cases, mounting the CD changer on the rear deck, just behind the rear seat is a good, safe place. In fact, you can facilitate your installation by adding plugs to the connections and sticking a small sheet of Velcro to the bottom so it can be easily removed. Surprisingly enough, the Velcro holds the player firmly in place against the carpeted rear deck. Mounting it this way also keeps vibrations from bumping the CD off track when it is playing.

Multiple speaker systems require a cross-over network to direct sound frequencies to the appropriate speakers. You certainly don't want your tweeters woofing or your woofers tweeting so a network such as this one from Alpine should be included in any "killer" stereo system.

Amplifiers

If you want to add on to your existing stereo, one of the first items to consider is a power increase. There are several manufacturers who make graphic equalizers and amplifiers that bump up the power of a stock stereo receiver. Any time the power of a stereo system is increased, so is maximum potential volume, but more importantly, sound at all volume levels is crisper and cleaner.

The one problem with an add-on amp or equalizer is finding a good mounting location. An equalizer with built-in amp may require an installer to mount it on the dash. Additional amplifiers depend on the receiver to which they'll be connected. Simple systems use a single amp for rear speakers while multiple amplifiers can be used to power each speaker section.

Using multiple amplifiers will soon become a problem because of the lack of space. Using such a system will undoubtedly require a custom installation, since multiple amps will more than likely require a cross-over network to channel the individual signals to each speaker. By this point, you're getting into pretty technical territory, and unless you have superior electronic knowledge, you might want to go with a pro for the installation. Multiple

Nothing is ever simple and stereo systems are no exception. Multiple speakers can't deliver good sound unless the stereo system has sufficient power to drive them. Rated at 120 watts, this Pioneer amp has what it takes to supply "supercharged" sound.

amplifiers however, do make a tremendous sound system that will put your Camaro in the ranks of competition vehicles.

Amp Selection—Choosing the right amplifier may be a little more difficult than you expect. The first consideration in choosing an amp is to determine what power levels your speakers require and the size and space limitations. For the most part, you need to look at the total wattage an amplifier puts out continuously to each channel. Many amplifiers will also state the amount of distortion at that wattage and obviously, the one with the least amount of distortion is the better unit.

Mounting Amps—In late-model Camaros, a good place to mount your amplifiers is in the rear hatch area. All third-generation Camaros have a deep well in the hatch and that space can hold a couple of mid-sized amplifiers. If you want a custom installation, a small enclosure can be made just behind the rear seat where you can mount and display your amps if you plan on entering sound competitions.

Speakers

Many professional stereo installers and competitors say that you can't appreciate a good stereo if you have poor quality speakers. In fact, a good set of speakers will make your stock system sound better as well. Unfortunately, the Camaro is limited as far as speaker locations are concerned and if you want additional speakers, they will have to be placed in custom made enclosures.

Stock Speakers—To begin with, the stock speaker locations on 1982-and-later Camaros are up front under the dash, and in the rear pillars directly under the hangar hooks. The front speakers are 4x6 units while the rears are 6 x 9's. If you are planning on replacing the stock units, make sure the new ones are specifically made for Camaros since

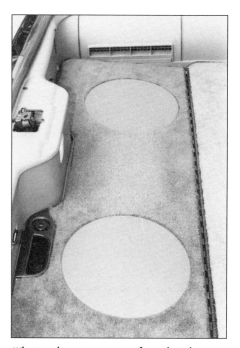

What good is a super stereo if you don't have speakers that can handle ear drum-splitting volume? Two 12-inch woofers, mounted in the package area can handle enough bass volume to scare small children and shake the foundations of mobile homes. Careful where you play it though—many communities have enacted laws to discourage high-volume stereo playing from your car.

All 1982-and-later Camaros incorporate two 6x9 inch speakers in the rear. Replacing the stock speakers with a high-quality three-way system, such as this one from Pioneer, deliver maximum sound at minimal expense while requiring no cutting of interior components.

133

Here's an example of maximizing the limited rear space in a Camaro. Alpine amps are tucked into the hatch, covered by this custom, flip-top lid that houses a pair of sub-woofers.

they need to be of a low-profile design.

Replacing the stock speakers is very simple and using a set of higher quality aftermarket units will enhance the sound of your system. If you use an installation kit with a harness adapter, then you don't have to wire the speakers and you can use the factory wires to hook up the aftermarket units.

If you use the stock locations, you will have a simple four-speaker system. For the best sound from a four-speaker system, replace the factory speakers with two- or three-way speakers. Pioneer manufactures two-way 4x6, low-profile speakers for the front dash; they're made specifically for Camaro applications. These can be used with additional Pioneer three-way 6x9 speakers for the rear pillar area. Any additional speakers will require some kind of custom installation.

Additional Speakers—If you do want to add more speakers, two of the more common areas to add them are on the doors and in the rear hatch area. The doors on late-model Camaros are large enough to hold several speakers. Many enthusiasts use a good five- or six-inch mid-range speaker with a tweeter mounted towards the front of the door, just under the armrest and near the front kick panels.

There are two ways of mounting speakers here. The best and simplest is to use a pre-manufactured, surface-mount enclosure that mounts to the outer side of the door in front of the door panel. This requires a minimum number of bolt holes through the door and the speakers mount inside this enclosure for better sound. You can use a number of different grilles or covers to match your Camaro's interior.

The other way to mount door speakers is to cut large holes so the speakers can be recessed in the door itself. However, there is very limited space inside the door, so depending on the speaker's size, you may have no choice other than a surface mount. If you don't have power windows and plan on installing speakers in the door, you should leave the installation to an experienced stereo installer.

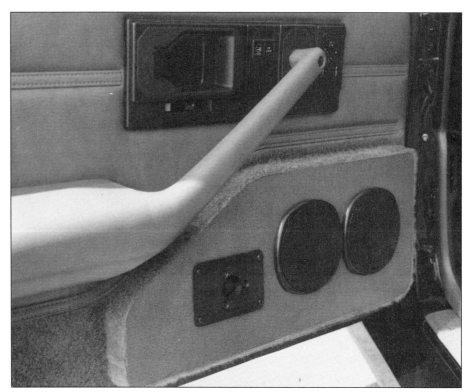

Although it can be tricky to install, a door-mounted enclosure is another way to add multiple speakers for a superior sound.

Under the dash, Camaros have a 4x6 inch speaker. These Alpine two-way speakers can be easily mounted in place of the stock units for superior sound quality.

SECURITY SYSTEMS

Now that you have a nice stereo system in your Camaro, as well as thousands of dollars of high performance equipment, you may want to keep it around. Third-generation Camaros were once the most-often stolen vehicle in the United States. The advent of the Pass Key system, adapted from the Corvette, on 1989- and-later Camaros, has reduced vehicle theft somewhat. The Pass Key system includes a special module on the ignition key. This module contains a specially encoded code and if the key inserted in the ignition switch doesn't contain the proper code, the vehicle can't be started. At the same time the Pass Key system was being implemented in the Camaro, a anti-theft system for factory stereos was also being used. This system locks the factory stereo and disables it so it can't be operated. This anti-theft device was only available on the optional Delco/Bose systems. However, your aftermarket stereo is still at risk, and if you own a pre-1989 Camaro, you

As noted, another suitable area for installing additional speakers is the rear hatch. This area serves as the best position for some sub-woofers for rich, deep bass sounds. Adding speakers here can be simple and will also allow you to be very creative. In many instances, the entire rear hatch area has been turned into a giant enclosure that is flush with the rear seat and that houses several 10" or 12" sub-woofer speakers.

You can add as many speakers back here as your budget allows, but for a good, simple, inexpensive installation, your best bet is to cut out an enclosure from a 1/2" to 1" thick piece of inexpensive wood or particle board. The board must be 40" by 15" and you can cut two speaker holes to flush-mount two 8", 10" or 12" subwoofers. Your local auto interior shop should have suitable carpeting that matches your Camaro's interior and once you have carpeted the enclosure and mounted your speakers, you'll have tremendous bass for your system at a minimal cost.

If you want to avoid the burden of cutting out an enclosure for your Camaro, there is a neat box made by Kicker that houses two, 8" subwoofers and two tweeters that fits snugly in the rear hatch area. This box was designed to fit in the rear of Camaros and provides great sounds for a relatively low cost.

One thing that you should remember if you are attempting a multiple speaker system, is that you will need appropriate power to operate these speakers and you may also need a cross-over unit to channel the specific frequencies to high, mid-range and bass speakers.

Look ma, no steering wheel. Grant offers a unique anti-theft system that includes a steering wheel with a special adapter that allows the wheel to be removed and replaced in a matter of seconds. With no way to steer a vehicle, it's much less attractive to thieves.

Camaros are hot cars in more than one way; thieves love them. Therefore, installation of a security system is advisable. This system from Vehicle Security Electronics has a multi-function remote that can operate doors, windows, starter and perform various other functions.

have no security at all. Investing in a security system will definitely give you peace of mind as well as minimizing the possibility of theft.

While no system is foolproof, there are some security systems better than others, and it certainly will add some protection. The best systems employ a computer to determine when a break-in has occurred and sound a loud alarm. But like everything else, high tech isn't cheap, so a less sophisticated system might be in order

One of the most sophisticated alarm systems is the Derringer 2 by Vehicle Security Electronics Inc. It uses the computer to determine where the break-in occurred and flashes a code to tell you whether an attempt was made at a door, hood, rear hatch or other area.

Systems such as these can have a full range of options, including the capability of rolling up your car's windows and locking the doors—all with just the flick of a switch on a wireless remote. Some even have functions which will allow you to roll down the windows and pop open the rear hatch by remote. Higher priced systems can also include a pager unit and will allow your remote to turn on your stereo system, start the engine, open your doors if they are solenoid operated, and even open your garage door.

The important thing about selecting a good security system is to choose one in your price range that does everything you want it to do. Secondly, make sure that you purchase your unit from a reputable and experienced installer who can mount and wire the system correctly. The installer should also offer warrantees on both the unit and the installation.

Another good feature on an alarm system is a small remote control. There are a lot of systems that work great but have a big, bulky remote that is uncomfortable when you carry it around in your pocket.

Security systems can be either *active* or *passive*. Passive systems automatically arm the vehicle as you leave it, while active systems must be armed manually, usually with a remote transmitter. Some insurance companies will give discounts for having an alarm system, so it pays to check with home.

Whatever type of system you decide on, the important thing is to make a conscious effort to care for your Camaro responsibly. This goes for driving as well. If you've worked your way through all of the chapters in this book, and followed our advice, then you should have the hottest Camaro in your neighborhood. But remember to use good judgment and drive safely at any speed. Good luck!

"This is your security system speaking." Some systems like this one from Viper, have a voice module that tells someone when they are too close to the vehicle. It's like having a built-in mother-in-law.

INDEX

A
Adams, Herb, 108, 114
AFR (Air Flow Research) heads, 69-70
Aftermarket
 brake kit, 119
 computers, 26-27
 lower control arms, 122
 mufflers, 44-45
 rear ends, 105
 stock differentials, 105
 thermostats, 54
 transmissions, 100-101
Aftermarket PROMs, 22-23
 replacing, 23
 retaining stock calibrations, 23
Aftermarket springs, 115-116
Air cleaners, 54-55
 dual-snorkel, 55
 inlets into the engine, 55
Air flow enhancer, 57
Air foils, 58-59
 TPI throttle body, 59
Air inlets, 55
Airflow, key to, 69
Alignment bars, 74
Amplifiers, audio, 132-133
 mounting of, 133
 selection of, 133
Anti-roll bars, 116-117
 increasing size of, 116-117
Arms
 custom lower control, 115
 rocker, 73-74
 stock rear control, 117
Arms, lower control, 120, 122
 aftermarket choices, 122
Assembly Line Diagnostic Link (ALDL), 18
Automatic transmissions, 94

B
Bars, alignment, 74
Batteries, fully charged, 37
Berlinetta, 7
Billet wheels, 111
Brake kit, aftermarket, 119
Brakes, 120-121
Bushings, 117-119
 nylon, 119
 polyurethane, 117
 replacement, 118-119
 ride & noise changes, 119
 rubber, 117-119
 urethane, 118-119

C
California Air Resources Board exemptions, 22, 27
Calipers, replacement, 118
Cam profiles, flat-tappet, 76
Camaro, 1982 year, 2-3
 pace car edition, 2
 Z28 package, 23
Camaro, 1983 year, 4
Camaro, 1984 year, 4-5
Camaro, 1985 year, 5
 IROC feature, 5
Camaro, 1986 year, 5-6
 1LE, 6
 technical highlights, 6
Camaro, 1987 year, 6-7
 technical highlights, 6-7
Camaro, 1988 year, 7-8
 technical highlights, 7-8
Camaro, 1989 year, 8-9
 technical highlights, 8-9
Camaro, 1990 year, 9-10
 1LE, 9-10
 technical highlights, 10
Camaro, 1991 year, 10-11
Camaro, 1992 year, 11-13
 technical highlights, 12-13
Camaro
 25th Anniversary, 100
 fourth generation, 1, 13
 history of, 1-13
Camaro LT, 7
Camber, negative and positive, 109
Camshafts, 74-77
 and carbureted engines, 76
 duration of, 74-75
 facts & figures, 75-76
 lift of, 75
 selection guidelines of, 74-75, 76-77
Carbureted engines & camshafts, 76
Carbureted intake manifolds, 56-57
Carbureted systems, 55-57
Carburetors, Holley, 56
Carburetors and computers, 20-21
Carburetors, Rochester, 56
 LG4 Quad, 56
Cast wheels, 111
Caster, 108-109
Cat back system, 40, 41-42
Catalytic converters, 41-43
Center high mounted stop light (CHMSL), 5
Centrifugal superchargers, 82
Chevrolet, 10-11
Chips (computer), 22, 24-25
Clutches, 101-102
Coil springs, 112-115
Coil systems, 37
Commemorative Edition pace cars, 2
Compact disc players and changers, 131-132
 mounting of, 131-132
Competition belts, 128-129
Composite wheels, 111
Computerized engine control systems, 15-27
Computers
 aftermarket, 26-27
 and carburetors, 20-21
 designed for racing, 27
 PC compatible, 19
Coolant Temperature Sensors, 17
Cornering, 116
Corvette heads, 69
Crankshaft pulley, underdrive, 18
Cross Fire Injection, 4
Crossfire, 31-32
Cylinder heads, 68-69, 69
 & valvetrain modifications, 67-77
 shopping for, 70-71

D
Detonation sensor, 83
Diacom software, 19-20
Diacom systems, 18
Differentials, stock, 103-104
 aftermarket for, 105
 Dana 44, 104
 Ford 9-inch, 105
 modifications, 104
Distributor cap, 33
Drivetrains, 93-105
Dumping, ECM computer, 22

E
Electronic Control Module (ECM), 15-16
 learning ability of, 22
Electronic engine management systems, 15-27
Electronic fuel injection, 20
Emissions-legal nitrous oxide installation, 89-91
Engine control systems, computerized, 15-27
Engines,
 tuned port injection engines, 87-88
EPA regulations, 21
Escort Endurance Challenge, 9

137

Exhaust Gas Oxygen (EGO) Sensor, 16
Exhaust manifolds/headers, 45-48
    choosing headers, 46-48
Exhaust system, installing a performance, 49-51
Exhaust system design, stock, 40-41
Exhaust systems, 39-51
    cat back, 40, 41-42
    short primary pipes, 40
Extrude Honing, 60

F

Federal Clean Air Act
    fines, 41
    violations of, 48
Forced induction, 79-91
Ford 9-inch rear ends, 105
Four-tube headers, 47
Fuel injection parts, replacement, 27
Fuel pressure regulators, 61-62

G

GM ECM's, learning ability of, 22

H

Handling, subjective evaluation, 111
Head swapping, 69-70
Headers
    legality of, 47-48
    materials used to make, 48
    tubing sizes of, 47
Headers, choosing, 46-48
Heads
    Corvette, 69
    small-block, 70
    street, 72
    V6, 71
Heads, AFR, 69-70
Heads, Corvette, 69
Heads, cylinder, 68-69
Heat ranges, 33-36
    and octane, 34-36
    of spark plugs, 33-34
Heritage appearance packages, 11-12
High Energy Ignition System (HEI), 7
High flow mufflers, 43-45
High Output option engines (H.O.), 4
History of Camaro, 1-13
Holley carburetors, 56
Horsepower, key to, 69

I

Idle Air Control (IAC), 17
Ignition system, killer, 30
Ignition systems, 29-37
    aftermarket, 31
Induction, forced, 36-37, 79-91
Induction systems, 53-65
Inductive crossfire, 32
Injector nozzles, 62-63
Installing
    Grant GT performance steering wheel, 127
    nitrous oxide systems, 89-91
    Paxton supercharger, 84-85
    performance exhaust system, 49-51
    power chip, 24-25
    Richmond 5-speed transmission, 98-99
    subframe connectors, 123
Intake manifolds, 56, 57
Interior modifications, 125-136
IROC (International Race of Champions), 5, 11

K

Knock sensor, 17

L

Law & catalytic converters, 41-43
Legality of headers, 47-48
Lift, camshafts, 75
Limp home mode, 17-18
Lingenfelter, John, 61
Lower control arms, 120, 122
    custom, 115

M

Manifold Absolute Pressure (MAP) Sensors, 17
Manifold Air Temperature (MAT) Sensors, 16, 65
Manifolds
    carbureted intake, 56-57
    low-profile aluminum intake, 56
    TPI, 60
Manual transmissions, 97-102
Marcel, 102
Mass Air Flow (MAF) Sensors, 16, 17
Modular wheels, 111
MSD, 31
Mufflers,
    aftermarket, 44-45
    high flow, 43-45

N

Nitrous oxide
    defined, 87
    injection, 87-88
Nitrous oxide systems, 86
    installing, 88, 89-91
Nozzles, injector, 62-63
Nyliner inserts, 119

O

Octane and heat ranges, 34-36

P

Pace car
    edition (1982), 2
    Indianapolis 500 (1982), 2
Panhard rods, 117, 119-120
Pass Key
    System, 135
    theft deterrent system defined, 8
Paxton supercharger, installing a, 84-85
Performance modifications, 21-22
    of the computer, 21
Plenums, 60-61
Plug gap, 33
Plus System, The, 110
Pocket porting, 70
Port matching and TPI runners, 59
Porting, 70
Power losses, parasitic, 55
Power secrets, TPI, 64-65
Prom Paq switching systems, 21
PROMs, aftermarket, 22-23
    replacing, 23
PROMs (Programmable Read Only Memory), 16
    add-on devices, 26
    special computers & conversions, 23-26
Pulleys, underdrive, 55

R

Rear discs, 120-121
Rear ends, 102-105
    Ford 9-inch, 105
Receivers, audio, 130-131
Regular Production Option (RPO), 3
Regulators, fuel pressure, 61-62
Richmond 5-speed transmission 98-99
Ride height, adjusting, 115-116
    aftermarket springs, 115-116
    dropped spindles, 116
    lowering, 115

Rochester, QuadraJets, 56
Rocker arms, 73-74
Roll bars, 129
Roll stiffness, increasing, 116
Roots-type superchargers, 81-82
Rubber shim, 115
Runners, TPI, 59

S

SCCA Escort Endurance Challenge, 9
SCCA (Sports Car Club of America), 9
Seatbelts, 128-129
    competition belts, 128-129
Seats, 126-129
Security systems, 135-136
    passive or active, 136
Sensors, 16
    Idle Air Control (IAC), 17
    Knock Sensors, 17
    Limp Home mode, 17-18
    Manifold Absolute Pressure (MAP) Sensors, 17
    Manifold Air Temperature (MAT) Sensors, 16
    Mass Air Flow (MAF) Sensors, 17
    PROM, 16
    Throttle Position Sensors (TPS), 16-17
Servo assembly, 700-R4, 95
Sheet metal spider, 74
Shift improving kits, 96
Shifters, 96
Shim, rubber, 115
Shock absorbers, 117
    adjustable, 117
    proper dampening, 117
Spark plugs, 33
    recommendations for, 33-34
    gaps, 33
    gaps and forced induction, 36-37
    heat ranges of, 33-36
    wires for, 30
Spark timing, 36
Speakers, 133-135
    additional, 134-135
    stock, 133-134
Speedometer, realistic, 126
Spider, sheet metal, 74
Spring load, 115
Spring rate, 115
Spring sag, 115
Springs, progressive and linear, 116
Springs, valve, 73
Steering wheel, installing a Grant GT, 127
Steering wheels, 126

Stereo systems, 129-131
Stock differentials, 103-104
Stock exhaust system design, 40-41
Stoichiometric, 16, 62
Subframe connectors, 121
    installing, 123
Supercharger, installing a Paxton, 84-85
Supercharger types, 81-82
    centrifugal, 82
    roots, 81-82
Superchargers, complimentary components, 82-83
    computer chips, 82-83
    high performance ignition, 83
    increases in spark intensity, 83
    manual control devices, 83
    timing control units, 83
Superchargers, low profile, 80
Supercharging, 81-83
Suspension, 107-123
Swapping transmissions, 101
Systems components, 16-18
    Coolant Temperature Sensors, 17
    Electronic Control Module (ECM), 16
    Exhaust Gas Oxygen (EGO)

T

Terminals, distributor cap, 33
Thermostats, aftermarket, 54
Third generation, 1-13
Throttle Body Injection (TBI), 57
Throttle Position Sensors (TPS), 16-17
Timing control units, 83
Tire size, increasing, 110
Tires, 110-111
    selection of, 110-111
    The Plus System, 110
Toe, 109-110
Toe, zero, 110
Toe-in, 109
Toe-out, 110
Torque converters, 96-97
Torx bolts, 62
TPI
    engines, 26
    inlets, 55
    manifolds, 60
    power secrets, 64-65
    runners, 59
    runners and manifold, 58
    runners and port matching, 59
    throttle body air foils, 59
    throttle bodies, 61
TPI systems, installation of, 20

Tranny Swap, 98-99
Transmission reprogramming kits, 95
Transmissions
    aftermarket, 100-101
    installing a Richmond 5-speed, 98-99
    racing 100,
    swapping, 101
Transmissions, automatic, 94-97
    700-R4, 94-97
Tri-Y headers, 46
Trouble codes, 18-20
    third-generation Camaros and, 19
Tubing sizes of headers, 47
Tuned port injection (TPI), 3, 6, 58-63, 88
Turbo lag, 83
Turbo systems, 86
Turbocharging, 83-87
    disadvantages of, 86-87

U

Underdrive crankshaft pulley, 18

V

V6 heads, 71
Valve
    job, 73
    springs, 73
Valves, options for purchasing, 72
Valves & valvetrains, 71-74
Valvetrain modifications, 67-77
Volumetric efficiency (VE), 79

W

Weight transfer, 116
Wheels, 111-112
    alignment of, 108-110
    backspacing of, 111
    billet, 111, 114
    cast, 111
    composite, 111
    custom, 112
    modular, 111
    selection of, 111-112

Z

Z28 package (1982), 2-3
Zerk fittings, 119

OTHER BOOKS FROM HPBOOKS AUTOMOTIVE

HANDBOOKS
Auto Electrical Handbook: 0-89586-238-7
Auto Upholstery & Interiors: 1-55788-265-7
Brake Handbook: 0-89586-232-8
Car Builder's Handbook: 1-55788-278-9
Street Rodder's Handbook: 0-89586-369-3
Turbo Hydra-matic 350 Handbook: 0-89586-051-1
Welder's Handbook: 1-55788-264-9

BODYWORK & PAINTING
Automotive Detailing: 1-55788-288-6
Automotive Paint Handbook: 1-55788-291-6
Fiberglass & Composite Materials: 1-55788-239-8
Metal Fabricator's Handbook: 0-89586-870-9
Paint & Body Handbook: 1-55788-082-4
Sheet Metal Handbook: 0-89586-757-5

INDUCTION
Holley 4150: 0-89586-047-3
Holley Carburetors, Manifolds & Fuel Injection: 1-55788-052-2
Rochester Carburetors: 0-89586-301-4
Turbochargers: 0-89586-135-6
Weber Carburetors: 0-89586-377-4

PERFORMANCE
Aerodynamics For Racing & Performance Cars: 1-55788-267-3
Baja Bugs & Buggies: 0-89586-186-0
Big-Block Chevy Performance: 1-55788-216-9
Big Block Mopar Performance: 1-55788-302-5
Bracket Racing: 1-55788-266-5
Brake Systems: 1-55788-281-9
Camaro Performance: 1-55788-057-3
Chassis Engineering: 1-55788-055-7
Chevrolet Power: 1-55788-087-5
Ford Windsor Small-Block Performance: 1-55788-323-8
Honda/Acura Performance: 1-55788-324-6
High Performance Hardware: 1-55788-304-1
How to Build Tri-Five Chevy Trucks ('55-'57): 1-55788-285-1
How to Hot Rod Big-Block Chevys:0-912656-04-2
How to Hot Rod Small-Block Chevys:0-912656-06-9
How to Hot Rod Small-Block Mopar Engines: 0-89586-479-7
How to Hot Rod VW Engines:0-912656-03-4
How to Make Your Car Handle:0-912656-46-8
John Lingenfelter: Modifying Small-Block Chevy: 1-55788-238-X
Mustang 5.0 Projects: 1-55788-275-4

Mustang Performance ('79–'93): 1-55788-193-6
Mustang Performance 2 ('79–'93): 1-55788-202-9
1001 High Performance Tech Tips: 1-55788-199-5
Performance Ignition Systems: 1-55788-306-8
Performance Wheels & Tires: 1-55788-286-X
Race Car Engineering & Mechanics: 1-55788-064-6
Small-Block Chevy Performance: 1-55788-253-3

ENGINE REBUILDING
Engine Builder's Handbook: 1-55788-245-2
Rebuild Air-Cooled VW Engines: 0-89586-225-5
Rebuild Big-Block Chevy Engines: 0-89586-175-5
Rebuild Big-Block Ford Engines: 0-89586-070-8
Rebuild Big-Block Mopar Engines: 1-55788-190-1
Rebuild Ford V-8 Engines: 0-89586-036-8
Rebuild Small-Block Chevy Engines: 1-55788-029-8
Rebuild Small-Block Ford Engines:0-912656-89-1
Rebuild Small-Block Mopar Engines: 0-89586-128-3

RESTORATION, MAINTENANCE, REPAIR
Camaro Owner's Handbook ('67–'81): 1-55788-301-7
Camaro Restoration Handbook ('67–'81): 0-89586-375-8
Classic Car Restorer's Handbook: 1-55788-194-4
Corvette Weekend Projects ('68–'82): 1-55788-218-5
Mustang Restoration Handbook('64 1/2–'70): 0-89586-402-9
Mustang Weekend Projects ('64–'67): 1-55788-230-4
Mustang Weekend Projects 2 ('68–'70): 1-55788-256-8
Tri-Five Chevy Owner's ('55–'57): 1-55788-285-1

GENERAL REFERENCE
Auto Math:1-55788-020-4
Fabulous Funny Cars: 1-55788-069-7
Guide to GM Muscle Cars: 1-55788-003-4
Stock Cars!: 1-55788-308-4

MARINE
Big-Block Chevy Marine Performance: 1-55788-297-5

HPBOOKS ARE AVAILABLE AT BOOK AND SPECIALTY RETAILERS OR TO ORDER CALL: 1-800-788-6262, ext. 1

HPBooks
A division of Penguin Putnam Inc.
375 Hudson Street
New York, NY 10014